# SpringerBriefs in Psychology

More information about this series at http://www.springer.com/series/10143

Lisa M. PytlikZillig • Myiah J. Hutchens
Peter Muhlberger • Frank J. Gonzalez
Alan J. Tomkins

# Deliberative Public Engagement with Science

## An Empirical Investigation

Lisa M. PytlikZillig
Public Policy Center
University of Nebraska
Lincoln, NE, USA

Peter Muhlberger
Public Policy Center
University of Nebraska
Lincoln, NE, USA

Alan J. Tomkins
Public Policy Center
University of Nebraska
Lincoln, NE, USA

Myiah J. Hutchens
Edward R. Murrow College of
Communication
Washington State University
Pullman, WA, USA

Frank J. Gonzalez
School of Government & Public Policy
University of Arizona
Tuscon, AZ, USA

Additional material to this book can be downloaded from http://extras.springer.com.

ISSN 2192-8363          ISSN 2192-8371   (electronic)
SpringerBriefs in Psychology
ISBN 978-3-319-78159-4          ISBN 978-3-319-78160-0   (eBook)
https://doi.org/10.1007/978-3-319-78160-0

Library of Congress Control Number: 2018936754

This Springer imprint is published by the registered company Springer International Publishing AG part
of Springer Nature.
The registered company address is: Gewerbestrasse 11, 6330 Cham, Switzerland

# Acknowledgments

The research and data dissemination activities were supported by the National Science Foundation (NSF) under grants #0965465 and #1623805 from the SciSIP (Science of Science and Innovation Policy) program. Any opinions, findings, and conclusions are those of the authors and do not necessarily reflect the views of the SciSIP program or NSF.

We are grateful to the many individuals who contributed to the present research either as colleagues, student participants, and/or student assistants in facilitating the process of the engagements and/or the research on the engagements. Colleagues who provided particularly significant input included Yuris Dzenis, Jack Morris, Ted Pardy, and Joe Turner of the University of Nebraska-Lincoln. Graduate students who were involved in the project were Ryan Anderson, Tim Collins, Frank Gonzalez, Joe Hamm, Jeremy Hanson, Becky Harris, Ashley Johnson, Chris Kimbrough, Ryan Lowry, Matt Morehouse, Jayme Nieman, Peibei Sun, Shiyuan Wang, and Deadric Williams. Student assistants who participated in conducting the research with support from an NSF REU (Research Experience for Undergraduates) Award 0965465 were Hina Acharya, Whitney Aurand, Mark Batt, Rob Broderick, Dorothy Chen, Jamie DeTour, Addison Fairchild, Chris German, Kayla Kumm, Jessica Loke, Jen McCarty, Macey Morgan, Brock Nelsen, and Jarred Vogel.

# Contents

# About the Authors

**Frank J. Gonzalez** is an Assistant Professor in the School of Government and Public Policy at the University of Arizona. He received his PhD in Political Science from the University of Nebraska-Lincoln.

**Myiah J. Hutchens** is an Assistant Professor at the Edward R. Murrow College of Communication at Washington State University in Pullman, Washington. She received her PhD from The Ohio State University in Communication.

**Peter Muhlberger** is a Research Fellow at the Public Policy Center at the University of Nebraska. He previously served as a Program Director in Cyber Social Science and Political Science at the National Science Foundation. He received his PhD in Political Science from the University of Michigan.

**Lisa M. PytlikZillig** is a Research Associate Professor at the University of Nebraska Public Policy Center, Assistant Director of the University of Nebraska-Lincoln Social and Behavioral Sciences Research Consortium, and has a courtesy appointment in the Department of Psychology at University of Nebraska-Lincoln. She received her PhD in Personality and Social Psychology at the University of Nebraska-Lincoln.

**Alan J. Tomkins** worked on this research in his roles as the Director of the Public Policy Center at the University of Nebraska and Professor at the University of Nebraska-Lincoln Law/Psychology Program. He has since retired from the university and is emeritus director and professor, and he has joined the National Science Foundation's Directorate for Social, Behavioral, and Economic Sciences (NSF/SBE). He received his PhD in Social Psychology and a JD from Washington University.

# Chapter 1
# The Big Picture

**Abstract** The purpose of this book is to share some results and the data from four studies in which we used experimental procedures to manipulate key features of deliberative public engagement to study the impacts in the context of deliberations about nanotechnology. In this chapter, we discuss the purpose of this book, which is to advance science of public engagement, and the overarching question motivating our research: What public engagement methods work for what purposes and why? We also briefly review existing prior work related to our overarching goal and question and introduce the contents of the rest of the book.

**Keywords** Science of public engagement · Deliberative engagement · Science and technology studies · Nanotechnology · Big data

## 1.1 Introduction

Some of us remember the time before widespread Internet access, when instead of watching YouTube or Facebooking, we watched a preset schedule of Saturday morning cartoons. One such cartoon, *The Jetsons*, featured a futuristic family that lived a seemingly amazing life—populated by then-imaginary inventions such as video phones, housecleaning robots, and flying cars.[1] Now, of course, video phones are old news and have exceeded Jetson-inspired expectations: instead of mounting them on the wall, you can carry them in your pocket. Robots increasingly Roomba our carpets, Robomow our lawns, and have begun to patrol our

---

[1] http://www.smithsonianmag.com/history/50-years-of-the-jetsons-why-the-show-still-matters-43459669/

**Electronic supplementary material**: The online version of this chapter (https://doi.org/10.1007/978-3-319-78160-0_1) contains supplementary material, which is available to authorized users.

© The Author(s) 2018
L. M. PytlikZillig et al., *Deliberative Public Engagement with Science*,
SpringerBriefs in Psychology, https://doi.org/10.1007/978-3-319-78160-0_1

shopping malls.[2] And, as a final step toward the Jetsonian life, news outlets recently have been abuzz with commentary about the development of flying cars.[3]

At the same time as new technological developments bring futuristic dreams to life and widen imaginable opportunities, they also often result in unanticipated new problems. George Jetson had to grapple with pizza for breakfast when his food dispenser malfunctioned and with a robot co-worker that was trying to steal his job. Today, there is increasing interest in "robot-proof" jobs. Meanwhile, cyberbullying, sexting, and social isolation are examples of problems attributed to widespread smartphone use. Texting and driving has resulted in a troubling new reason for car crashes, encouraging authorities to consider the potential use of "textalyzer" technology to detect when drivers are illegally texting just before a crash.[4] Others worry about the dramatic increase in data collected on everyday citizens, the potential rise of a pervasive surveillance society, use of big data to manipulate people, and the unknown effects of nanoparticles that can easily cross the blood-brain barrier.

Given the potential for negative—or at least controversial—effects of new technologies upon the societies in which various publics must live, what could be more democratic than promoting public involvement in decisions about those new technologies? Unless, of course, it turns out that public involvement, which can sometimes be costly, is ineffective, unnecessary, or actually makes things worse. Some have suggested this may be the case (e.g., Sunstein, 2000, 2002), but, for better or worse, public engagement with and about new technologies is happening all around us. Our interest in studying such public engagement—the topic of this book—is to learn how to design it for the better.

The research described in this book was funded by the National Science Foundation (NSF)[5] and aimed to begin to fill current gaps in the research on public engagement by applying certain social, psychological, and behavioral theories and experimental procedures. As we describe in Chap. 2, our project included five studies, four of which we present in this book.[6] The four studies described here involved more than 1000 college students as participants, and all four studies focused on the same topic and context. Thus the studies resulted in a wealth of quantitative and qualitative data collected at multiple time points and provide a unique opportunity to see which results replicate across studies.

Our work was motivated by a desire to better understand how, when, and why public engagement might work to achieve different purposes. It also reflects a largely untapped role that social scientists might play in the area of responsible

---

[2] http://www.npr.org/2017/04/26/525675196/robot-security-guards-coming-to-shopping-malls

[3] http://www.npr.org/sections/alltechconsidered/2017/04/25/525540611/flying-cars-are-still-coming-should-we-believe-the-hype

[4] http://www.npr.org/sections/alltechconsidered/2017/04/27/525729013/textalyzer-aims-to-curb-distracted-driving-but-what-about-privacy

[5] Research and data dissemination is funded by NSF #0965465 and #1623805. Any opinions, findings, and conclusions are those of the authors and do not necessarily reflect the views of the National Science Foundation (NSF).

[6] Study 1 data, our pilot data, was prioritized last for release and is currently not included in the full release of data. Researchers wishing to use our data are welcome to do so as long as they cite it appropriately.

research and innovation (Macnaghten, Kearnes, & Wynne, 2005). Prior scholars have noted that social scientists are needed to help in the design phase of the technology (Doubleday, 2007; Evans & Kotchetkova, 2009). By facilitating public engagement, social scientists can help technologists revise their work so that it not only "works" in a technical sense but also in social sense, so that it doesn't, for example, suffer the polarized fate of genetically modified foods in Europe (Gaskell, Bauer, Durant, & Allum, 1999; Marris, 2015; Webler & Tuler, 2010).

However, our view is that social scientists are also needed to take the lead in theorizing, researching, and advancing the *science of public engagement*. The field could use some bona fide "public engagement psychologists," as well as "public engagement political scientists" and "public engagement communication" researchers. Work in education might provide a model for the new field or set of fields we envision. Understanding and promoting positive educational outcomes are not the goal of a single field. Rather, diverse scholars are involved in advancing education-relevant goals, including those who study educational psychology, educational policy, and educational administration. There is a need for similarly diverse groups of scholars to work from different angles to focus specifically upon how to promote engagement-related outcomes in and across specific contexts.

In our studies we worked from a psychological perspective to begin to envision and demonstrate what a science of public engagement might look like. The aims of this book are to tell the story of our experience; share our measures, methods, and data; describe some of our findings; and ultimately (hopefully) provoke and inspire more studies advancing the science of public engagement. We hope our story will embolden and facilitate additional attempts to apply rigorous experimental procedures to public engagement contexts and that our overviewed studies provide exemplars for future efforts. By providing links to our detailed methods, materials, and measures and reams of quantitative and qualitative data, we hope to foster additional analyses and findings and maybe also to provide materials useful for training others who aspire to be public engagement scientists. Indeed, our rich data likely have further insights to reveal to researchers with a variety of interests. The data also reveal, in sometimes humbling ways, the struggles we encountered in conducting our experiments. We hope lessons from our struggles can enhance future studies of public engagement strategies used in different contexts and for varied purposes.

In light of these aims, we use the rest of this chapter to provide a brief overview of the existing public engagement literature and gaps that motivated our research. We also discuss some social and psychological theories potentially applicable to the development of science of public engagement, and provide an overview of the rest of the book.

## 1.2  Motivating Questions and Gaps

Public engagement is claimed to have numerous benefits (Fiorino, 1990). Proponents claim it is "the right thing to do" (Petts, 2008) and that it will result in better and more publicly acceptable policies. Those policies, they say, will take into account more viewpoints, while the engagement activities simultaneously improve citizenship capacities (Selin et al., 2016) and social capital (Webler, Kastenholz, & Renn, 1995). Imagine, if you will, George Jetson getting to have a say about the design of his robot co-workers. George might suggest a precautionary "no-job-stealing" algorithm be installed before the robot begins working. As a result of George's engagement, not only does policy and technology development improve, but George himself learns about a new technology and its pros and cons, as well as learning about and gaining appreciation of others' views and honing skills needed to express his own views—ultimately improving our democracy, one George at a time.

Other more skeptical writers provide contrary claims that public engagement might actually have harmful effects. What if George Jetson and his human colleagues fail to imagine important effects of new technologies? How useful is their input then? What if the engagement incites polarization and conflict among participants instead of fomenting forward-moving consensus (Kahan, 2012; Schkade, Sunstein, & Hastie, 2007; Sunstein, 2002)? George and his bottom-line-focused boss may have very different ideas about how the robot co-worker should be developed. Some writers also argue engagement may hand the powerful even more power (Benhabib, 2002; Hickerson & Gastil, 2008) or cause citizens to disengage rather than engage (Hibbing & Theiss-Morse, 2002). After all, if George finds inventors and regulators catering to his boss's concerns rather than his, what reason does he have to engage in the future?

Despite the negative possibilities, a number of democracies seem to agree that the public should be engaged around technology and policy decisions. In the latter part of the twentieth century, the Netherlands began developing and using a procedure called constructive technology assessment (CTA) as a means to include more stakeholder perspectives and ensure that social values were taken into account earlier in technology development (Rip & Robinson, 2013; Wilsdon & Willis, 2004). More recently, in the USA, public participation was touted as a key feature of the Obama Administration's Open Government Initiative. Now, as Internet giants and other organizations increasingly make online experimentation and other research a part of their everyday operations, international guidelines have been released, encouraging public deliberation aimed at defining the appropriate ethical boundaries for such big data social research.[7]

Calls and support for public engagement have become so widespread that some have claimed we are, for better or worse, in an "age of engagement" (Delgado, Kjølberg, & Wickson, 2011). Certainly we are in an age of *calls* for public engagement, which suggests that public engagement, and what it really achieves, should be given more attention. What measurable good and/or harm does public engagement do? How, when, and why does it do so? Unfortunately, despite all the enthusiasm for public engagement, as well as some pointed doubts and criticisms, the empirical research on public engagement is still in its infancy. Especially few are the number of controlled experiments that might elucidate microprocesses and psychological factors that operate during public engagements and perhaps shed light on conflicting outcomes from prior work.

---

[7] http://www.oecd-ilibrary.org/science-and-technology/research-ethics-and-new-forms-of-data-for-social-and-economic-research_5jln7vnpxs32-en

As described elsewhere (PytlikZillig & Tomkins, 2011), we began our exploration of public engagement by considering the evidence base for some of the claims made about the benefits of public engagement. Very quickly, we realized that results from studies of public engagement were highly variable. Thorough academic reviews of such variable outcomes are provided by others (Delli Carpini, Cook, & Jacobs, 2004; Mendelberg, 2002; Ryfe, 2005), and we do not repeat them here. However it is useful to consider an illustrative example, such as the outcomes of engagements around planning for potential influenza pandemics.

The mention of the word "pandemic" probably fills some people's heads with visions of Ebola and SARS and others with the words "it won't happen to me." Both responses are or perhaps should be frightening. In this context, Garrett, Vawter, Prehn, DeBruin, and Gervais (2009, p. 18) "strongly urge government officials and policymakers to facilitate robust public engagement on key issues in pandemic ethics," arguing that "[i]nformed public perspectives can help improve pandemic policies, promote trust and enhance cooperation."

Yet, an evaluation of six pandemic engagement projects conducted in the USA was much more cautious in its conclusions about the effectiveness of such engagements (Public Policy Center, 2010). On the positive side, the report noted the engagement events did appear to result in overall changes in opinions about the types of social values that should be weighed during a pandemic. This suggests that the effort to engage and inform people had some effect, resulting in changes in attitudes due to the engagement activities. Yet, on the negative side, *agreement* on the values underlying people's opinions did not increase. Further, relating to the hoped-for "informed public perspectives," the report noted that participant knowledge did generally increase, but also that, "given the relatively low post process scores across states, we cannot conclude that participants were well informed" as they gave their input (pp. 12–13). Results relating to the promotion of institutional trust were also mixed. Some of the evaluated projects showed increases in trust in some institutions, others showed decreases in trust in some institutions, and still others showed mixed or no changes in trust.

Such varied results, which are common across studies of public engagement, give hope that public engagement *can* have positive effects but also underscore that positive impacts are not certain. Different outcomes can, will, and do occur under different conditions; but there is little clarity regarding which conditions, features, or contexts are responsible for the differences. This leads to a recognized problem in the public engagement literature: the lack of clarity around "how to enable effective involvement (i.e., which mechanism to use, and how) in any particular situation," (Rowe & Frewer, 2005, p. 252) or, phrased another way, "*which* forms, features, and conditions of public engagement are optimal for *what* purposes and *why*" (PytlikZillig & Tomkins, 2011, p. 198). We were interested in such questions because they not only have implications for theory development, they also are essential for providing practitioners with direction for designing "effective" public engagement in different situations. Thus, in the next sections, we break down our overarching question into its component parts (what works, for what purposes, and why) and describe the state of the prior research pertaining to each.

### *1.2.1    What Works? Delineating Important Public Engagement Types and Variables*

Answering the question of "*what* works" requires defining what public engagement is or is not, as well as identifying different dimensions or types of engagement. Although some narrower definitions have been offered (Litva et al., 2002), currently, just about any interaction with the public that is even tangentially policy-related seems to count as public engagement. Public engagement includes activities ranging from idea marketing and museum exhibits, to focus groups and national surveys, to community-based participatory research, to citizen juries and deliberations (Rowe & Frewer, 2005).

Attempts to define and organize the myriad of public engagement activities have included placing them on a "ladder" to reflect the amount of power they afford public, ranging from total citizen control to public manipulation (Arnstein, 1969; see also Pretty, 1995). Other suggested distinctions include purpose of the engagement (Glass, 1979; Rosener, 1975), structure of activities (Glass, 1979), public acceptability (Nelkin & Pollak, 1979), types of participants (Cornwall, 2008; Fung, 2006), and direction of information flow to and/or from the public (Rowe & Frewer, 2005). Some have also noted that variation occurs both between *and within* different types of public engagement. For example, Carman et al. (2013) focused only on "deliberative" engagement mechanisms and noted that these can vary in their recruitment methods, number of participants, use of face-to-face versus online modes of interaction, use of different resources such as educational materials and accessible experts, and length and number of sessions.

Despite the considerable work done to organize and name all the variations, it's still not really clear what factors, dimensions, or characteristics of public engagement are most worthwhile to study. Almost two decades ago, Chess and Purcell (1999) noted that "typologies" of engagement mechanisms do not seem to consistently correspond to different outcomes—casting doubt on how useful it is to simply compare different broad "types" of engagement (e.g., surveys vs. deliberations vs. focus groups). Later, Rowe and Frewer (2005) advised that researchers should prioritize the study of design variables most likely, from a theoretical and empirical standpoint, to impact the *effectiveness* of the engagement activity. Of course, theorizing about effectiveness also requires defining what counts as success (or as positive outcomes) when it comes to public engagement, a topic which we turn to next.

### *1.2.2    For What Purposes? Assessing Engagement Effectiveness and Success*

The public engagement literature has also given quite a bit of attention to defining and organizing criteria and measures for the success of public engagements. Some of these criteria come from the arguments for or against public engagement, which then have been arranged in categories of success criteria. For example, Webler et al. (1995) proposed two categories: fairness and competency criteria. *Fairness* of an engagement activity is judged by how acceptable the activity is to the public, its inclusiveness of affected stakeholders as participants, the extent to which processes are equitable and transparent, and so on. *Competency* of an engagement refers to how well and efficiently it achieves its purposes, whether those purposes are to educate and inform, to gather the full range of viewpoints on an issue, to build trust, or something else.

Rowe and colleagues (Rowe & Frewer, 2000; Rowe, Horlick-Jones, Walls, Poortinga, & Pidgeon, 2008) somewhat similarly categorized criteria for judging the success of public participation activities into acceptance criteria or process criteria. *Acceptance* criteria include whether the participants are representative of the affected public and whether the event occurs early in decision-making, in a transparent and unbiased manner. *Process* criteria include having well-defined tasks, highly accessible and appropriately thorough and unbiased resources, appropriately structured decision-making processes, and cost-effective methods.

In an attempt to align common effectiveness criteria with workflow processes associated with designing and implementing public engagements, PytlikZillig and Tomkins (2011) suggested that categories of *information criteria* (e.g., is the information balanced, complete, accurate) and *representation criteria* (e.g., are all relevant stakeholders included) are associated with preparing for the engagement, *process* and *acceptance criteria* (e.g., are the appropriate processes implemented effectively and found to be acceptable by participants) are associated with implementing the engagement, and *outcome criteria* (e.g., did the engagement achieve its goals) are associated with the purposes and hope-for functions of the engagement.

While these classes of criteria provide useful overviews of everything about an engagement that might be judged and evaluated, there are at least a couple[8] of problems with using the criteria classes to advance theory and research. Most important to the work we present in this book, the classes are too broad to readily lend themselves to the application and testing of specific theories. Each class of criteria contains varied constructs, and each construct may need its own theoretical and empirical account. Very few evaluative frameworks have focused on tying specific engagement mechanisms to specific outcomes (but see Beierle, 1998's evaluation framework based on social goals). Research and theory might be advanced more quickly if effectiveness components were identified and organized in a manner that allowed for the application of specific theories to specific processes and outcomes and contexts.

---

[8] Due to our study design and space constraints, we will not be able to deal much with a second perhaps even more significant problem than discussed here, which is that most all of the outcome criteria are focused solely on the publics who are engaged and not on the experts or policymakers who also may be engaged or may have contracted the engagement. As researchers and practitioners increasingly seek out alternatives to "deficit models" of engagement, it is becoming more important to attend, not only to how publics are impacted by engagements but how policymakers', policies', and technologists' understandings, trust, and so on are also impacted (Eaton, Burnham, Hinrichs, & Selfa, 2017).

## 1.2.3    In What Contexts and Why? From Comparison
##             to Causation

The importance of context for public engagement has been extolled in the political science and STS (science, technology, and society) literatures (Delgado et al., 2011; Delli Carpini et al., 2004). In some ways, however, context seems to be a scapegoat for "inconsistent results." That is, the argument goes like this: Context must matter, because studies that analyze, compare, and even pit one type of engagement against another, in various contexts, find inconsistent results. Indeed, some studies, mostly conducted in health policy contexts, have compared deliberation, education-only, and measurement-only control groups or survey, interview, or discussion procedures. These studies often find greater change in knowledge and/or attitudes when deliberative methods are used instead of other methods (Abelson et al., 2003; Barabas, 2004; Carman et al., 2014; De Vries et al., 2010; Kim et al., 2011). However, other studies, such as Denver, Hands, and Jones's (1995) study of deliberative poll participants in the UK, find no change in knowledge or attitudes, and yet others suggest deliberation may facilitate the biased strengthening of pre-existing attitudes (Kahan, 2012; Sunstein, 2002). Even within our single program of research, which used highly similar methods, measures, and participants, we found inconsistent results from one study to another, as we describe in later chapters.

For the most part, it is still unclear whether the differences in results that come from diverse studies in the field are due to process differences such as variation in the operationalization of "deliberation" or whether studies are truly illustrating effects due to the context in which the processes are used or whether the effects are simply unstable and difficult to consistently achieve. Regardless, let's assume context does matter: "context" still doesn't provide a very informative explanation for different results. Findings that effects vary across studies and contexts beg for an answer to the question: Why? And "why" questions in turn beg for analyses of "how" and the use of methods that can test causal processes.

Experimental studies of public engagement that use random assignment and control groups, especially those that connect features-processes-outcomes, are increasing but still relatively rare (Carman et al., 2013; Friess & Eilders, 2015). Such studies can, however, be very fruitful and enlightening. For example, relating to the knowledge increases commonly found during engagements, a deliberative experiment by Muhlberger and Weber (2006) found that knowledge gains were more likely due to reading the materials, with no additional knowledge gains attributable to the deliberative discussion. In a later study, Esterling, Neblo, and Lazer (2011) used experimental methods to compare deliberative discussion that included online discussion with one's political representatives, with an information-only group, or a true control group (receiving no information). In contrast to the Muhlberger and Weber study, their methods found deliberative discussion participants gained more knowledge than either the control or information-only groups. But they also showed the increased knowledge was likely due to seeking out policy-relevant information outside of the experiment. This suggests their participants were motivated to appear informed about the issues in front of their representatives. Furthermore, it provides a potential explanation for why the findings may have differed from the Muhlberger and Weber study, based on time allowed for exploration of information outside the context of the engagement activities.

Studies that connect features-processes-outcomes can help clarify why similar but different designs have varied effects and be used to build theories that help to predict and explain when public engagement will be effective for different purposes. As we detail in Chap. 2, the framework we used to guide our methods also aspired to make features-processes-outcomes connections while at the same time examining moderators of such connections which might inform theory development.

## 1.3   Advancing the Theoretical and Empirical Bases of a Science of Public Engagement

### 1.3.1   The Current State of Theory

In the above review, we have identified a number of frameworks and exemplar studies but virtually no overarching "theories" of public engagement. This is, in part, due to the already discussed problems the field has faced when it comes to conceptualizing the "what" and "for what purposes" of public engagement. As Rowe and Frewer (2005) noted, "Given the sheer number of mechanisms available for engaging the public and also the confusion as to what each does and does not entail, and how each differs from the others, it is unsurprising that no significant theory has emerged as to what mechanism to use in what circumstance to enable effective engagement" (pp. 259–260). It is easy to extend Rowe and Frewer's observation about public engagement mechanisms (see also Chess & Purcell, 1999) to public engagement outcomes: Given the diverse outcomes public engagement is expected to achieve, it is not surprising that no single theory has emerged to cover all of them.

This is not to say that public engagement research is atheoretical or that theories are not applied within public engagement research. In some ways, the problem is that there are many, many—perhaps too many—theories, and these theories are not yet well-organized in a manner that optimally serves the advancement of a science of public engagement. Theories applied to public engagement range from the relatively encompassing perspective of deliberative democracy (see Gutmann & Thompson, 2004, for a review) to more narrow and specific theories such as Barabas' (2004) theory of opinion updating during deliberative discussions. The theory of affective intelligence (Marcus, Neuman, & MacKuen, 2000) also has been proposed and used to explain when individuals will engage versus disengage and how they will interact with political information under different conditions, including when they will engage in an open-minded versus defensive manner (MacKuen, Wolak, Keele, & Marcus, 2010). More broadly, agency theory (Muhlberger, 2005; Muhlberger & PytlikZillig, 2016) has been proposed as an overarching framework that might guide research on public engagement, taking into account both individual and group level, rational and nonrational, and psychological and sociological forces on public attitudes and beliefs. However, perhaps due to a lack of coordination across the specific areas and domains in which public engagement takes place, for the most part, current theories have not been integrated into more broadly useful theories that advance science of public engagement per se. Such theories also have not been integrated with other social psychological and cognitive theories in a manner that could help to advance those existing theories, or, in the case of the highly integrative agency theory, they have not been widely applied.

### *1.3.2  Moving Forward*

How might we move forward to develop more useful and integrative theories of public engagement that more effectively advance a science of public engagement? In our prior work (PytlikZillig & Tomkins, 2011), we have suggested that it would be useful to (1) analyze public engagements by inventorying their commonly varied features and hoped for outcomes; (2) broadly consider a variety of existing theories from a wide number of fields ranging from psychology to political science, to communication, and more; (3) narrow our focus and use experimental methods to carefully and systematically vary a subset of engagement features that (a) are purported as most important to achieving public engagement outcomes and (b) have strong connections to other existing theories; and (4), across multiple studies, systematically measure the impacts of those experimentally varied features on both the outcomes and potential explanatory mediators while also exploring potential moderators of the effects. This is the approach we therefore applied in the present work.

## 1.4  Focus and Overview of the Rest of this Book

Earlier in this chapter, we provided an overview of the many features of public engagement and the many outcomes public engagements are expected to achieve. Of course no single research program could examine all such features and outcomes. For our studies, we narrowed our focus to deliberative public engagements primarily because of the interests of our team members, each of whom had research and/or practical experience in contexts using deliberative methods.

*Deliberation* has been broadly defined as a type of thinking where people take in and weigh diverse information to form and justify their opinions (Gerber, Bächtiger, Fiket, Steenbergen, & Steiner, 2014; Gundersen, 1995; Lindeman, 2002). Democratic deliberative theory (Bohman, 2000; Chambers, 2003; Coleman & Gotze, 2001; Fishkin & Luskin, 2005; Gastil & Levine, 2005; Habermas, 1990) likewise purports that voters should first be thinkers and discussants who weigh the reasons for their choices. Thus, deliberative engagements are designed around the idea that the best decisions are ones that are critically evaluated and well-reasoned, and deliberative engagement stresses the need to consider diverse perspectives and weigh evidence in terms of its quality and relevance.

Beyond the cognitive definition of deliberation, others have suggested additional criteria before an activity can be called "deliberative engagement." For example, it is common to require the social criterion of discussion with others or formal processes for creating rationales such as "problem analysis, criteria specification, and evaluation" (Gastil, 2000, p. 22). Thompson (2008) argues that deliberation requires the social criterion of some disagreement and decision criteria involving a collective decision that will bind all group members (see also Parker, 2003). Even so, Thompson notes that non-binding discussion, such as that occurs in many deliberative polls (Fishkin & Luskin, 2005), may be seen as relevant preparation for democratic decisions, and studies of such processes therefore have relevance to understanding the effectiveness of deliberative methods.

Our research did not aim to compare "deliberation" to "non-deliberation," so much as it aimed to focus on specific features of deliberative engagement and their effects, alone or in combination. For this purpose, it is less important that we define deliberation exactly than it is that we identify some of the features that arguably are part of deliberation, which can then be operationalized and subjected to experimental manipulation. To facilitate effective deliberation, it is commonly argued that deliberative engagements need to include features such as *discussion*, high-quality *information*, participants who engage in *critical thinking* during exposure to *diverse opinions*, and with the help of effective *facilitation* (Bohman, 2000; Chambers, 2003; Coleman & Gotze, 2001; Fishkin & Luskin, 2005; Gastil & Levine, 2005; Habermas, 1990). We thus focused on varying these features in our studies. Meanwhile, some of the most commonly lauded and hoped-for outcomes of deliberative engagements include *informed (knowledgeable) opinions*, *attitude changes*, increased *trust in institutions*, and *acceptance of resulting policies*. These variables thus became some of our primary outcome variables. In Chap. 2 we describe how we operationalized such variables in our manipulations and measures, along with providing links to our materials and detailed method reports.

In Chaps. 3, 4, and 5, we provide examples of analyses focused on testing theories and exploring potentially important mediators and moderators that might advance various theories. The variables we focused upon are relevant to a wide number of theories from other fields that could be (but have yet to be) usefully tested in the context of public engagement—furthering both public engagement theory and the original theories. Such theories suggest a number of mechanisms or mediators by which features of engagement might impact outcomes. For example, theories and research from educational psychology are very relevant to public engagement learning outcomes. Yet certain constructs commonly examined in educational contexts, like individual differences in *intrinsic and extrinsic motivation* (McGregor & Elliott, 2002; Midgley, 2014; Pintrich, 2004) or use of different modes or types of *cognitive engagement* (Chin & Brown, 2000; Dinsmore & Alexander, 2012), still are not commonly examined in the context of public engagement. Furthermore, despite the emphasis of most public engagement activities on "informed opinions," little research has focused on what is meant by "informed." Informedness indicated by *subjective knowledge* often has a relatively low correlation with *objective knowledge* (Carlson, Vincent, Hardesty, & Bearden, 2008), and little theoretical (or empirical) guidance exists regarding whether engagement practitioners should focus on one versus the other.

In Chap. 3, we explicitly examine different types of cognitive-affective engagement as potential mediators of the effects of our experimental conditions upon changes in knowledge. We examine both how our conditions impacted these various types of engagement, as well as how different forms of engagement related to gains in subjective and objective knowledge. Our findings suggest self-reports of careful and conscientious engagement are especially important if the goal is to increase objective or subjective knowledge and that this conscientious engagement can be encouraged during deliberations via the instructions given to the participants.

Chapter 4 focuses on attitude formation and change and potential moderators of the effects of different engagement features on attitude outcomes. Like knowledge, attitudes toward the topics of the public engagement are commonly examined during public engagement evaluations. Typically such studies focus on whether individual attitudes change their attitudes or come to exhibit certain features like certainty or coherence (Gastil & Dillard, 1999). In addition, some studies examine attitudes at the group level, to determine the conditions under which attitudes held by a group of individuals show overall mean change, or come to consensus, exhibit polarization (Schkade et al., 2007), or move toward single-peakedness (Farrar et al., 2010). The literature on attitude formation and change is voluminous and includes reference to theories such as the theory of cognitive dissonance (Festinger, 1957), theories of motivated cognition (Kunda, 1990), the elaboration likelihood model and the meta-cognitive model (Petty & Brinol, 2010), and so on. These theories, although too often not explicitly mentioned, are relevant to prior studies of attitudes during public engagement, including prior experimental studies (e.g., see Baccaro, Bächtiger, & Deville, 2016). In Chap. 4, we discuss the application of such theories and explore whether our experimental conditions relate to attitude changes at the individual and group levels and to the observed variation in participant attitudes. By also examining moderators of some of our effects, we move toward determining the reliability of the relationships between our experimental manipulations and attitude outcomes.

In Chap. 5, we explore whether certain variables may operate via mediation and moderation processes simultaneously. We do this while discussing an undertheorized and underinvestigated outcome: policy acceptance. Drawing from existing theories related to policy acceptance and support, procedural fairness, and legitimacy, we propose that certain engagement features or processes and perceptions of engagement processes may simultaneously impact mediating and moderating variables which, at times, may work against one another to hide main effects of the experimental condition. Using correlation and multiple regression analyses, we demonstrate that, despite the lack of main effects of our experimental conditions upon policy acceptance, our experimentally varied features of deliberative engagement are impacting mediators and moderators in ways that have implications for advancing theory and practice.

Finally, in Chap. 6, we summarize some of the key lessons that we learned from our efforts on these studies over the years. In our studies, we attempted to reach beyond evaluations of public engagement, to focus upon theory development, by creating methods, materials, and measures that draw broadly from a number of diverse and relevant theoretical perspectives. We discuss how successful these efforts were and some of the drawbacks and benefits of our approach. Hopefully, our frank assessments will spark continued conversations regarding other ways in which the development of theories of public engagement might take place and how empirical research on public engagement might be expanded in the future.

# References

Abelson, J., Eyles, J., McLeod, C. B., Collins, P., McMullan, C., & Forest, P.-G. (2003). Does deliberation make a difference? Results from a citizens panel study of health goals priority setting. *Health Policy, 66*(1), 95–106.

Arnstein, S. R. (1969). A ladder of citizen participation. *American Institution of Planners Journal, 35*, 216–224.

Baccaro, L., Bächtiger, A., & Deville, M. (2016). Small differences that matter: The impact of discussion modalities on deliberative outcomes. *British Journal of Political Science, 46*(03), 551–566.

Barabas, J. (2004). How deliberation affects policy opinions. *American Political Science Review, 98*, 687–701.

Beierle, T. C. (1998). *Public participation in environmental decisions: An evaluation framework using social goals* (Vol. Discussion paper 99–06). Washington, DC: Resources for the Future. Retrieved from http://ageconsearch.umn.edu/bitstream/10497/1/dp990006.pdf.

Benhabib, S. (2002). *The claims of culture: Equality and diversity in the global era*. Princeton, NJ: Princeton University Press.

Bohman, J. (2000). *Public deliberation: Pluralism, complexity, and democracy*. Cambridge, MA: MIT press.

Carlson, J. P., Vincent, L. H., Hardesty, D. M., & Bearden, W. O. (2008). Objective and subjective knowledge relationships: A quantitative analysis of consumer research findings. *Journal of Consumer Research, 35*(5), 864–876.

Carman, K., Heeringa, J., Heil, S., Garfinkel, S., Windham, A., Gilmore, D., ... Pathak-Sen, E. (2013). The use of public deliberation in eliciting public input: Findings from a literature review. AHRQ Publication No. 13-EHC070-EF (pp. Prepared by the American Institutes for Research under Contract No. 290-2010-00005.). Rockville, MD: Agency for Healthcare Research and Quality. Retrieved from www.effectivehealthcare.ahrq.gov.

Carman, K., Maurer, M., Mallery, C., Wang, G., Garfinkel, S., Richmond, J., ... Fratto, A. (2014). Community forum deliberative methods demonstration: Evaluating effectiveness and eliciting public views on use of evidence. AHRQ Publication No. 14(15)-EHC007-EF (pp. 1–316). Prepared by the American Institutes for Research under Contract No. 290-2010-00005). Rockville, MD: Agency for Healthcare Research and Quality. Retrieved from www.effectivehealthcare.ahrq.gov.

Chambers, S. (2003). Deliberative democratic theory. *Annual Review of Political Science, 6*(1), 307–326.

Chess, C., & Purcell, K. (1999). Public participation and the environment: Do we know what works? *Environmental Science & Technology, 33*(16), 2685–2692.

Chin, C., & Brown, D. E. (2000). Learning in science: A comparison of deep and surface approaches. *Journal of Research in Science Teaching, 37*, 109–138.

Coleman, S., & Gotze, J. (2001). *Bowling together: Online public engagement in policy delib- eration.* London, UK: Hansard Society. http://catedras.fsoc.uba.ar/rusailh/Unidad%207/ Coleman%20and%20Gotze%20Bowling%20Together,%20online%20public%20engage- ment%20in%20policy%20deliberation.pdf.

Cornwall, A. (2008). Unpacking 'participation': Models, meanings and practices. *Community Development Journal, 43*(3), 269–283.

De Vries, R., Stanczyk, A., Wall, I. F., Uhlmann, R., Damschroder, L. J., & Kim, S. Y. (2010). Assessing the quality of democratic deliberation: A case study of public deliberation on the ethics of surrogate consent for research. *Social Science & Medicine, 70*(12), 1896–1903.

Delgado, A., Kjølberg, K. L., & Wickson, F. (2011). Public engagement coming of age: From theory to practice in STS encounters with nanotechnology. *Public Understanding of Science, 20*(6), 826–845.

Delli Carpini, M. X., Cook, F. L., & Jacobs, L. R. (2004). Public deliberation, discursive participa- tion, and citizen engagement: A review of the empirical literature. *Annual Review of Political Science, 7*, 315–344.

Denver, D., Hands, G., & Jones, B. (1995). Fishkin and the deliberative opinion poll: Lessons from a study of the Granada 500 television program. *Political Communication, 12*(2), 147–156.

Dinsmore, D. L., & Alexander, P. A. (2012). A critical discussion of deep and surface processing: What it means, how it is measured, the role of context, and model specification. *Educational Psychology Review, 24*, 499–567.

Doubleday, R. (2007). Risk, public engagement and reflexivity: Alternative framings of the public dimensions of nanotechnology. *Health, Risk & Society, 9*(2), 211–227.

Eaton, W. M., Burnham, M., Hinrichs, C. C., & Selfa, T. (2017). Bioenergy experts and their imagined "obligatory publics" in the United States: Implications for public engagement and participation. *Energy Research & Social Science, 25*, 65–75.

Esterling, K. M., Neblo, M. A., & Lazer, D. M. (2011). Means, motive, and opportunity in becom- ing informed about politics: A deliberative field experiment with members of Congress and their constituents. *Public Opinion Quarterly, 75*, 483–503.

Evans, R., & Kotchetkova, I. (2009). Qualitative research and deliberative methods: Promise or peril? *Qualitative Research, 9*(5), 625–643.

Farrar, C., Fishkin, J., Green, D., List, C., Luskin, R., & Paluck, E. (2010). Disaggregating delib- eration's effects: An experiment within a deliberative poll. *British Journal of Political Science, 40*, 333–347.

Festinger, L. (1957). *A theory of cognitive dissonance.* Evanston, IL: Row Peterson.

Fiorino, D. J. (1990). Citizen participation and environmental risk: A survey of institutional mech- anisms. *Science, Technology & Human Values, 15*(2), 226–243.

Fishkin, J. S., & Luskin, R. (2005). Experimenting with a democratic ideal: Deliberative polling and public opinion. *Acta Politica, 40*, 284–298.

Friess, D., & Eilders, C. (2015). A systematic review of online deliberation research. *Policy & Internet, 7*(3), 319–339. Retrieved from https://doi.org/10.1002/poi3.95.

Fung, A. (2006). Varieties of participation in complex governance. *Public Administration Review, 66*(s1), 66–75.

Garrett, J. E., Vawter, D. E., Prehn, A. W., DeBruin, D. A., & Gervais, K. G. (2009). Listen! The value of public engagement in pandemic ethics. *The American Journal of Bioethics, 9*(11), 17–19. Retrieved from https://doi.org/10.1080/15265160903197663.

Gaskell, G., Bauer, M. W., Durant, J., & Allum, N. C. (1999). Worlds apart? The reception of genetically modified foods in Europe and the US. *Science, 285*(5426), 384–387.

Gastil, J. (2000). *By popular demand.* Berkeley, CA: University of California Press.

Gastil, J., & Dillard, J. P. (1999). Increasing political sophistication through public deliberation. *Political Communication, 16*(1), 3–23.

Gastil, J., & Levine, P. (2005). *The deliberative democracy handbook: Strategies for effective civic engagement in the twenty-first century.* San Francisco, CA: Jossey-Bass.

Gerber, M., Bächtiger, A., Fiket, I., Steenbergen, M., & Steiner, J. (2014). Deliberative and non-deliberative persuasion: Mechanisms of opinion formation in EuroPolis. *European Union Politics, 15*(3), 410–429.

Glass, J. J. (1979). Citizen participation in planning: The relationship between objectives and techniques. *Journal of the American Planning Association, 45*(2), 180–189.

Gundersen, A. G. (1995). *The environmental promise of democratic deliberation*. Madison, WI: University of Wisconsin Press.

Gutmann, A., & Thompson, D. (2004). *Why deliberative democracy?* Princeton, NJ: Princeton University Press.

Habermas, J. (1990). *Moral consciousness and communicative action*. Cambridge, MA: MIT Press.

Hibbing, J. R., & Theiss-Morse, E. (2002). *Stealth democracy: Americans' beliefs about how government should work*. New York: Cambridge University Press.

Hickerson, A., & Gastil, J. (2008). Assessing the difference critique of deliberation: Gender, emotion, and the jury experience. *Communication Theory, 18*(2), 281–303.

Kahan, D. M. (2012). *Ideology, motivated reasoning, and cognitive reflection: An experimental study*. Cultural Cognition Lab Working Paper No. 107; Yale Law School, Public Law Research Paper No. 272. Retrieved from http://ssrn.com/abstract=2182588 or https://doi.org/10.2139/ssrn.2182588.

Kim, S., Kim, H., Knopman, D. S., De Vries, R., Damschroder, L., & Appelbaum, P. (2011). Effect of public deliberation on attitudes toward surrogate consent for dementia research. *Neurology, 77*(24), 2097–2104.

Kunda, Z. (1990). The case for motivated reasoning. *Psychological Bulletin, 108*(3), 480–498.

Lindeman, M. (2002). Opinion quality and policy preferences in deliberative research. *Political decision making, deliberation and participation, 6*, 195–224.

Litva, A., Coast, J., Donovan, J., Eyles, J., Shepherd, M., Tacchi, J., … Morgan, K. (2002). The public is too subjective': Public involvement at different levels of health-care decision making. *Social Science & Medicine, 54*(12), 1825–1837.

MacKuen, M., Wolak, J., Keele, L., & Marcus, G. E. (2010). Civic engagements: Resolute partisanship or reflective deliberation. *American Journal of Political Science, 54*, 440–458.

Macnaghten, P., Kearnes, M. B., & Wynne, B. (2005). Nanotechnology, governance, and public deliberation: What role for the social sciences? *Science Communication, 27*(2), 268–291.

Marcus, G. E., Neuman, W. R., & MacKuen, M. (2000). *Affective intelligence and political judgment*. Chicago, IL: University of Chicago Press.

Marris, C. (2015). The construction of imaginaries of the public as a threat to synthetic biology. *Science as Culture, 24*(1), 83–98.

McGregor, H. A., & Elliott, A. J. (2002). Achievement goals as predictors of achievement-relevant processes prior to task engagement. *Journal of Educational Psychology, 94*(2), 381–395.

Mendelberg, T. (2002). The deliberative citizen: Theory and evidence. In M. X. Delli Carpini, L. Huddy, & R. Shapiro (Eds.), *Research in micropolitics: Political decisionmaking, deliberation, and participation* (Vol. 6, pp. 151–193). Greenwich, CT: JAI Press.

Midgley, C. (2014). *Goals, goal structures, and patterns of adaptive learning*. New York, NY: Routledge.

Muhlberger, P. (2005). Human agency and the revitalization of the public sphere. *Political Communication, 22*(2), 163–178.

Muhlberger, P., & PytlikZillig, L. M. (2016). Agency theory: Toward a framework for research in the public's support for and understanding of science. In J. Goodwin (Ed.), *Confronting the challenges of public participation* (pp. 109–136). Ames, IA: Science Communication Project.

Muhlberger, P., & Weber, L. M. (2006). Lessons from the virtual Agoral project: The effects of agency, identity, information, and deliberation on political knowledge. *Journal of Public Deliberation, 2*(1), 1–39.

Nelkin, D., & Pollak, M. (1979). Public participation in technological decisions: Reality or grand illusion? *Technology Review, 81*(8), 54–64.

Parker, W. (2003). *Teaching democracy: Unity and diversity in public life.* New York, NY: Teachers College Press.

Petts, J. (2008). Public engagement to build trust: False hopes? *Journal of Risk Research, 11*(6), 821–835.

Petty, R. E., & Brinol, P. (2010). Attitude change. In R. F. Baumeister & E. J. Finkel (Eds.), *Advanced social psychology: The state of the science* (pp. 217–259). New York: Oxford University Press.

Pintrich, P. R. (2004). A conceptual framework for assessing motivation and self-regulated learning in college students. *Educational Psychology Review, 16*, 385–407. https://doi.org/10.1007/s10648-004-0006-x.

Pretty, J. N. (1995). Participatory learning for sustainable agriculture. *World Development, 23*(8), 1247–1263.

Public Policy Center. (2010). *Evaluation of Public Engagement Demonstration Projects for Pandemic Influenza.* Lincoln, NE: University of Nebraska Public Policy Center (PPC). Retrieved from http://ppc.unl.edu/wp-content/uploads/2010/05/P5-Report-FINAL.pdf.

PytlikZillig, L. M., & Tomkins, A. J. (2011). Public engagement for informing science and technology policy: What do we know, what do we need to know, and how will we get there? *Review of Policy Research, 28*, 197–217.

Rip, A., & Robinson, D. R. (2013). Constructive technology assessment and the methodology of insertion. In N. Doorn, D. Schuurbiers, I. van de Poel, & M. E. Gorman (Eds.), *Early engagement and new technologies: Opening up the laboratory* (Vol. 16, pp. 37–53). Dordrecht, Netherlands: Springer.

Rosener, J. B. (1975). A cafeteria of techniques and critiques. *Public Management, 57*(12), 16–19.

Rowe, G., & Frewer, L. J. (2000). Public participation methods: A framework for evaluation. *Science, Technology, and Human Values, 25*, 3–29.

Rowe, G., & Frewer, L. J. (2005). A typology of public engagement mechanisms. *Science, Technology & Human Values, 30*(2), 251–290. Retrieved from http://search.ebscohost.com-library.unl.edu/login.aspx?direct=true&db=buh&AN=16483185&site=ehost-live&scope=site.

Rowe, G., Horlick-Jones, T., Walls, J., Poortinga, W., & Pidgeon, N. (2008). Analysis of a normative framework for evaluating public engagement exercises: Reliability, validity and limitations. *Public Understanding of Science, 17*, 414–441. https://doi.org/10.1177/0963662506075351. Retrieved from http://pus.sagepub.com/cgi/content/abstract/0963662506075351v1.

Ryfe, D. M. (2005). Does deliberative democracy work? *Annual Review of Political Science, 8*, 49–71.

Schkade, D., Sunstein, C. R., & Hastie, R. (2007). What happened on deliberation day? *California Law Review, 95*(3), 915–940. Retrieved from http://scholarship.law.berkeley.edu/californialawreview/vol95/iss3/6.

Selin, C., Rawlings, K. C., Ridder-Vignone, K. d., Sadowski, J., Allende, C. A., Gano, G., ... Guston, D. H. (2016). Experiments in engagement: Designing public engagement with science and technology for capacity building. *Public Understanding of Science, xx*(ahead of print). https://doi.org/10.1177/0963662515620970. Retrieved from http://journals.sagepub.com/doi/abs/10.1177/0963662515620970.

Sunstein, C. R. (2000). Deliberative trouble? Why groups go to extremes. *The Yale Law Journal, 110*(1), 71–119.

Sunstein, C. R. (2002). The law of group polarization. *Journal of Political Philosophy, 10*(2), 175–195.

Thompson, D. F. (2008). Deliberative democratic theory and empirical political science. *Annual Review of Political Science, 11*, 497–520.

Webler, T., Kastenholz, H., & Renn, O. (1995). Public participation in impact assessment: A social learning perspective. *Environmental Impact Assessment Review, 15*(5), 443–463.

Webler, T., & Tuler, S. P. (2010). Getting the engineering right is not always enough: Researching the human dimensions of the new energy technologies. *Energy Policy, 38*(6), 2690–2691.

Wilsdon, J., & Willis, R. (2004). *See-through science: Why public engagement needs to move upstream*. London, UK: Demos.

# Chapter 2
# Specific Methods

**Abstract** In this chapter we provide an overview of the experimental methods used in our four research studies. We describe the context for our studies and describe our rationale for examining our research questions in the context of the college student classroom. Then we compare and contrast the major features of our studies, including the participants, timing, materials, measures, and procedures for each study, and provide explanations for certain changes made between studies. Finally, we provide information about how readers can find our more detailed materials and methods, which also accompany our data for Studies 2–4.

**Keywords** Framework for the study of engagement · Engagement features · College students · Future scientists · Ethical, legal and social issues

## 2.1 Connecting Features, Processes, and Outcomes During Deliberative Discussions

As noted in Chap. 1, when we began our studies, the relative lack of experimental research on public engagement led us to try to begin to fill that gap. Our consideration of what approaches (*public engagement features*) work for what purposes (*outcomes*) and why (i.e., via *what processes or mediators*) resulted in a general framework and conceptual strategy we applied to present research (PytlikZillig & Tomkins, 2011). This strategy involves considering some of the features commonly used and recommended for public engagement and then broadly considering how a variety of social and psychological theories might clarify how, when, and why those features might lead to various outcomes. The broad and inclusive consideration of relevant theories drove the design of our experiments, in which

**Electronic supplementary material**: The online version of this chapter (https://doi.org/10.1007/978-3-319-78160-0_2) contains supplementary material, which is available to authorized users.

we attempted to experimentally vary several features, while measuring and assessing a larger number of potential outcomes, mediators, and moderators. By conducting similar but varied procedures in highly similar samples over time, we were able to examine which findings are robust and which vary even within the context of our relatively narrow inquiry: engaging science students in deliberations about nanotechnology.

Our detailed methods, including all measures and materials for each study, accompany the data sets that are available in the supplemental files to this book. Note that our Study 1 was conducted as a pilot in which we tested the incorporation of our experimental methods into the classroom setting. Several of our methods and measures were changed based on feedback from students at the end of that first semester. Due to the different nature of Study 1, and the need to focus our resources on sharing our best data, we do not include much discussion of Study 1 in this book, and we are not presenting Study 1 data here.

In this chapter, we first describe the contexts and methodological features that were constant across the remainder of the studies (Studies 2–5) and explain why we believe studies in the college student context are important. We then summarize key differences and similarities between studies in experimental conditions and outcome measures, along with providing some of the rationale for the changes we made across studies.

## 2.2   Our Context: Future Scientists Deliberating About Nanotechnology over Time

It is important to study theories relevant to public engagement in the specific context of deliberation around science and technology development and policy. Empirical findings from the lab or even from one deliberative discussion to another do not always easily generalize. Science, technology, and society (STS) scholars generally agree that public engagement should be context sensitive (Delgado, Kjølberg, & Wickson, 2011), making it important to examine the impact of design factors within specific, concrete contexts. As part of our strategy for connecting features, processes, and outcomes, we held context as constant as possible across our experiments. While our approach necessarily limits the generalizability of results, it provides a solid foundation for establishing the existence of internally valid and robust results within our chosen context, before extending to others. As the reader will see in Chaps. 3, 4, and 5, even with all the controls we employed, finding consistent effects was, nonetheless, no small feat. The contextual features held constant in our studies include the type of participants involved, the topics of deliberation, and the use of a longitudinal, repeated measures design.

### 2.2.1   *Participants: College Students in the College Science Classroom*

To facilitate our use of experimental methods, we worked within the constraints of the college classroom, by engaging consecutive semesters of students enrolled in a freshman-level biology for science majors course at the University of Nebraska-Lincoln (UNL). Table 2.1 describes the basic demographics of the students in each of our studies. As shown, an estimated 85–90% of the students who began the course participated in the study each semester. Those not participating may have either dropped out of the course or not consented to let us use their data. Across all studies, participants involved slightly more females than males and had an average age of 19–20 years. Students in Study 5, however, were slightly older and more varied in age. In each study, 70–90% of the students were in their first 2 years of college, and around 70–80% were science majors. About 40–50% of the students reported an affiliation with the Republican party, and the remainder of the students were approximately equally likely to affiliate as Democrats or as Independent/ Others, except in Study 2, in which there were proportionally more Independent/ Others. The fall semesters generally involved larger numbers of students than the spring semesters and a greater proportion of students in their first year of college.

There are at least three reasons why we believe this context is worthy of study. First, as noted by McAvoy and Hess (2013, p. 19), classrooms are "one of the most promising sites for teaching the skills and values necessary for deliberative demo-cratic life." US college students have often just achieved voting age, and most are just beginning a fuller participation in democracy. It is an arguably worthwhile endeavor to include more deliberative democracy in the classroom, and experiments in such contexts will facilitate understanding of optimal ways to do just that. Second, within the realm of STEM (science, technology, engineering, and mathematics) education, there has been a movement toward creating curricula that result in more

**Table 2.1** Descriptive comparison of participants across studies

|  | Study 2 | Study 3 | Study 4 | Study 5 |
|---|---|---|---|---|
| Total N participants in data set | 198 | 316 | 205 | 317 |
| **Estimated % females (based on self-reports)** | **53%** | **56%** | **57%** | **52%** |
| Average (standard deviation) age in years | 19 (2) | 19 (2) | 19 (2) | 20 (3) |
| Age range | 17–30 | 17–40 | 17–36 | 16–46 |
| **First year in college (%)** | **58%** | **40%** | **60%** | **43%** |
| **Second year in college (%)** | **28%** | **40%** | **27%** | **29%** |
| Third year in college (%) | 10% | 15% | 10% | 19% |
| Fourth year in college or beyond (%) | 4% | 5% | 4% | 7% |
| **Science majors** | **76%** | **79%** | **79%** | **68%** |
| **Republican** | **41%** | **45%** | **47%** | **46%** |
| Democrat | 24% | 27% | 28% | 24% |
| Independent/other | 35% | 28% | 26% | 22% |

well-rounded graduates who are not only experts in their fields but also able to work with interdisciplinary teams and think about the implications of the technologies that they may work with or develop and refine. Within biology education at UNL, discussion of a "New Biology" that focuses on interdisciplinary problem-solving and the application of science to solving societal problems makes our work applicable to the goals of that movement (Labov, Reid, & Yamamoto, 2010; National Research Council, 2009). As noted by the National Research Council's 2009 publication of "New Biology for the 21st Century: Ensuring the United States Leads the Coming Biology Revolution" (p. 10):

> Science and technology alone, of course, cannot solve all of our food, energy, environmental, and health problems. Political, social, economic, and many other factors have major roles to play in both setting and meeting goals in these areas. Indeed, increased collaboration between life scientists and social scientists is another exciting interface that has much to contribute to developing and implementing practical solutions.

Thus, work like ours is useful for introducing future scientists to the social science that is likely to impact the practical *usefulness* of their work as it intersects with the public and a variety of public viewpoints. Third, a very practical reason for our study of public engagement within the college classroom context is that it allowed us the use of experimental methods such as random assignment to conditions, increasing the internal validity of our findings in that context.

## 2.2.2  Discussion Topics: Nano-Biological Technologies and Human Enhancement

In each study, the deliberative activities focused on emerging and potential nanotechnologies. Because the activities took place in a biology course, we focused on technologies that involved biological or health applications, such as the use of nanotechnology for creating new nanomedicines or for human enhancement. We chose nanotechnology as a topic of deliberation because, at the time of our studies, governments were calling for and sometimes requiring public engagement around nanotechnology. For example, in 2003, the US twenty-first Century Nanotechnology Research and Development Act (P.L. 108–153) required public input and outreach as part of ensuring "that ethical, legal, environmental, and other appropriate societal concerns...are considered during the development of nanotechnology." Abroad, a government-commissioned report on nanotechnology by the Royal Society and Royal Academy of Engineering argued for widespread and early public involvement during nanotechnology development (Royal Society/RAE, 2004).

Table 2.2 shows some of the features of the background documents that varied between studies. The focus of the Study 2 document was nanogenomics. Between Study 2 and 3, the topics of the background document were expanded to discuss nanotechnology in general, as well as nanogenomics and nanomedicine, and the ethical, legal and social issues (ELSI) surrounding these technologies. In addition, between Study 3 and 4, revisions were made in response to student feedback that the

**Table 2.2**  Comparison of background documents across studies

|  | Study 2 | Study 3 | Study 4 | Study 5 |
|---|---|---|---|---|
| Topics | Nanogenomics how it is used where it is heading what people are saying References and links to other materials | Nanotechnology Nanomedicine Nanogenomics ELSI-relevant topics References and links to other materials | *Same as study 3 but included additional information about the risks of nanotechnology* | *Same as study 4 but only in NIF structure form, and strong and weak versions created* |
| Approx. words* | 2500 | 4500 | 4900 | 4900 |
| Structure | Topical | Topical and NIF (pro-con) versions | *Same as study 3* | NIF (pro-con) |
| Prompts | Prompts separate from reading | Prompts embedded in reading | *Same as study 3* | *Same as studies 3 and 4* |
| Formats | PDF with clickable links and printed copies available during A3 in class | Online A2 reading with printed copies available in class during A3 | *Same as study 3* | *Same as studies 3 and 4* |

Notes: ELSI refers to ethical, legal, and social issues. Prompts for deeper engagement used along with the background readings are described in the text

*The NIF (National Issues Forum)-formatted materials tended to be a bit longer due to repeating some of the claims for each of the opposing (pro-con) perspectives

document seemed biased positively toward nanotechnology, to include additional information and resources relevant to risks related to nanotechnology. Some changes were also made to the format of the documents and their integration with the experimentally varied prompts. In Study 2 the background information was a stand-alone, downloadable PDF document, and the prompts for engagement were presented separately. In the other studies, the information was presented as web page text with clickable links, and prompts for student responses were embedded in the information, with students instructed to stop and answer each prompt before continuing to read the next web page. Students were given a link to a downloadable PDF at the end of their reading assignment to refer to throughout the rest of the engagement activities.

## 2.2.3   Repeated Measures Longitudinal Design

Each of the studies also involved repeated measures administered over approximately 4–14 weeks of the semester. Table 2.3 shows the timing sequence of activities for each study. As shown, most of the study activities were organized into assignments for the students to complete. The assignments were required of all students, and student work was graded for completion and at times for effort or quality. Generally speaking, however, if students completed the work, they were given full credit. All students were required to complete and turn in the assignments. Students were given

**Table 2.3** Timing and common content of study activities (assignments)

| | Study 2 Spring 2011 | Study 3 Fall 2011 | Study 4 Spring 2012 | Study 5 Fall 2012 |
|---|---|---|---|---|
| Study, course, dates, and time | *January to May 14 weeks* | *August to December 12 weeks* | *January to May 4 weeks* | *August to December 7 weeks* |
| A1* | **Mid-January** | **Late-August** | **Mid-March** | **Late-Sept** |
| | Explained the study and obtained informed consent. Assessed variables such as demographics, individual differences in trust, interest in politics, political self-efficacy, motivation for engaging in politics, deliberative citizenship, and knowledge about nanotechnology | | | |
| **Lecture** | **Early-April** | **Early-Nov** | **Early-April** | **Early-Nov** |
| | One hour lecture reviewing evidence of science being a reflection and part of society, discussing reasons why science should be regulated, and considering how science should be regulated and the role of public input | | | |
| **Recitation** | Introduction to science and technology at the intersection of nanotechnology and biology. Assign A2 to be due by next recitation. Short introduction or video on nanotechnology | | | |
| A2 | **Early-April** | **Early-Nov** | **Early-April** | **Early-Nov** |
| | Readings about nanotechnology and nanogenomics, nanomedicine, and various ethical, legal, and social perspectives on each. Measures of knowledge, attitudes, engagement, and evaluation of materials | | | |
| A3 | **Mid-April** | **Mid-Nov** | **Mid-April** | **Mid-Nov** |
| | Deliberate on a number of imagined future scenarios illustrating ethical, legal, and social issues pertaining to the use of nanotechnologies. Complete measures of engagement and if relevant group process, as well as measures of attitudes | | | |
| A4 | **Mid-April** | **Mid-Nov** | **Mid-April** | **Mid-Nov** |
| | Post-measures to assess changes in knowledge, attitudes, individual differences, and other evaluations | | | |
| A5 | **Late-April** | | | |
| | Study 2 included some additional measures that students could complete for extra credit. In this assignment we piloted new prompts and administered additional personality assessments. There was no A5 in the other studies | | | |

Notes: *A = activity or assignment, of which there were four. One difference between studies included the timing of A1. A1 was administered at the beginning of the semester for Studies 2 and 3 and thus up to 10 weeks prior to the other activities. A1 was conducted nearer in time (within 1–2 weeks prior) to the other activities for Studies 4 and 5

two opportunities to provide or withhold research consent: prior to assignment 1 and during assignment 4. Final consent decisions made in assignment 4 were honored. If a student did not complete assignment 4, then their consent decision for assignment 1 was honored. If a student did not complete either assignment 1 or 4, their data was omitted from the study.

Assignment 1 (A1) was assigned as homework for students to complete outside of class. This homework included reading an introduction describing public engage-

**Table 2.4**  Scenarios used as part of assignment 3 (A3) to prompt deliberation of ethical, legal, and social issues (ELSI)

| Study 2 | Study 3 | Study 4 | Study 5 |
|---|---|---|---|
| Human memory | Human memory | Human memory | Human memory |
| Cystic fibrosis | Cystic fibrosis | Cystic fibrosis | Cystic fibrosis |
| Nutritious food | | Barless prisons | Head injuries |
| | | Healthy chip | Illness reduction |

ment and why it is important and which gave an overview of the public engagement activities that would take place as part of the course. As part of A1, students also were asked to complete measures of demographics, their attitudes toward and knowledge of nanotechnology, trust, and other individual differences. The time that elapsed between A1 and other activities did vary between studies, which may have affected whether and how much students were exposed to other sources of information between A1 and other assignments. In Studies 2 and 3, A1 was completed very early in the semester, up to 10 weeks prior to the rest of the activities. In Studies 4 and 5, A1 was completed approximately 1–2 weeks prior to the rest of the engagement activities. The remainder of the activities, however, were always completed near to the end of the semester, over the course of approximately 2–3 weeks.

Just prior to assignment 2 (A2), students were given a 50-min guest lecture during a regularly scheduled large-group meeting of their course. The lecture was delivered by a member of the research team and described the role of public engagement in science and research. Then, during the week following the large-audience format lecture, students attended a small-group recitation. At that session, a researcher and their regular recitation instructor introduced them to the public engagement activities that would be done as part of the course. This introduction allowed students a chance to ask questions about the purposes of the activities and the assignment requirements. A video also was shown to introduce students to nanotechnology and its applications and to pique their interest. The video used for this purpose was a TED talk video available on YouTube. Assignment 2 (A2) was then also completed as homework on students' own time. A2 included the readings about nanotechnology and the experimentally varied cognitive engagement prompts. In addition, students were asked to complete measures of attitudes, knowledge, and engagement and to evaluate the reading materials.

Unlike the other assignments, assignment 3 (A3) was almost always completed during the students' 1-h recitation.[1] During A3, students were given brief descriptions of imagined future scenarios and questions designed to prompt deliberation about ethical, legal, and social issues related to nanotechnology development. Table 2.4 lists the scenarios used across the different studies. These scenarios, which are available in the supplemental materials, were developed by the research team and/or inspired by or adapted from scenarios used by other teams conducting public

---

[1] In Study 3, there were more participants than usual, and we assigned a subgroup of students to pilot an online discussion condition. These students completed A3 online outside of class instead of in-class.

engagements around nanotechnology (e.g., Hamlett, Cobb, and Guston, 2008).[2] The students completed these deliberative activities during class, under different conditions (e.g., working alone or while discussing with their peers) as discussed in the next section. Students also completed additional measures of attitudes and engagement and when relevant completed measures of group processes.

Finally, immediately after finishing A3 in class, students were given a link to online assignment 4 (A4). As part of A4, students completed a variety of post-measures including reporting their final attitudes and completing knowledge assessments. In the next sections, we give additional detail on the conditions varied as part of the assignments and the measures administered during each phase (assignment).

## 2.3  What Works? Experimentally Varied Deliberative Engagement Features

Table 2.5 compares the experimental manipulations that were used across studies. As shown, in all of the studies, we varied the presence or absence of explicit prompts to think critically. In studies 2–4, we experimentally manipulated the presence or absence of peer discussion. In Study 5, we varied the construction of discussion groups to represent homogeneous or diverse attitudes, the inclusion of passive or active facilitators, and the characteristics of the background information provided. We also varied an introductory opinion question about the assignments to see if the question might be impacting student perceptions of the assignment. While the full details are given in the detailed methodological reports, here we give an overview of the conditions and our rationale for examining them.

### 2.3.1  Importance of Ethical, Legal, and Social Issues (ELSI) Topics in Science Education

In our pilot Study 1 and in Study 2, we noticed a tendency for the science students who served as our participants to doubt whether the public participation activities were beneficial to them and should be part of their biology course. Part of the problem was, as students told us, the assignments were too long and boring, due to the many survey measures we had included. Another part of the problem, however, seemed to be an expectation that the activities themselves should not be part of a "basic biology" course. In response to these views, in Studies 3 and 4, prior to engaging in any of the deliberation activities, we asked all students the following open-ended question:

---

[2] We were especially influenced by scenarios and information in the National Citizens' Technology Forum (2007) Human Enhancement, Identity, and Biology: NCTF background materials.

**Table 2.5** Experimentally varied engagement features in each study

|  | Study 2 | Study 3 | Study 4 | Study 5 |
|---|---|---|---|---|
| Importance of ELSI consideration in science education (A1) | *No students answered a question concerning their opinion of import* | *All students answered a question concerning their opinion of import* | *All students answered a question concerning their opinion of import* | Question Control question |
| Characteristics of background information (A2) | *All topical* | Topical NIF pro-con | Topical NIF pro-con | NIF stronger NIF weaker |
| Prompts for cognitive engagement (A2, A3) | Control Critical thinking Information organization | Control Critical thinking | Control Critical thinking | Control Critical thinking |
| Peer discussion (A3) | Present Absent | Present Absent Online* | Present Absent | Homogeneous+ Homogeneous- Heterogeneous |
| Discussion facilitation (A3) | *All groups accompanied by active facilitation* | *All groups accompanied by active facilitation* | *All groups accompanied by active facilitation* | Active Passive |

Note: Italicized text describes variables held constant within a study but which were varied in other studies. Non-italicized text describes the 2–3 experimental groups to which students were randomly assigned in a between-group design, as described in the text

*The online groups of respondents were not randomly assigned to that condition and thus may not be equivalent to the students in the present/absent peer discussion conditions. The data for the online students is included in the Study 3 data set but may be treated a pilot data for comparing similar tasks undertaken accompanied by in-class, online, or no peer discussion

> What do you think? In your opinion, how important is it that science students--including beginning science students such as you and your classmates--learn how to think about the ethical, legal and social issues (ELSI) pertaining to science? In 2-3 sentences, give your answer and a brief explanation of why you think as you do.

Interestingly, student open-ended responses to this question suggested largely positive views prior to engaging in the activities, which led us to wonder if we had changed student assessments of the activities by asking them their opinions in A1. Thus, in Study 5, rather than asking all students the ELSI importance question, we randomly assigned only one-half of all the students to answer the question. The other half of the students were not asked to reflect on the usefulness of the assignment. They instead were asked to give initial answers regarding the development of nanotechnology and its regulation.

## 2.3.2 Characteristics of the Background Information

To ensure high-quality materials that were accurate with respect to their depiction of nanotechnology, nanoscientists on our team assisted in finding or recommending source materials and reviewed our final readings for accuracy and appropriateness. In Study 2, all students read the same background document which was organized topically—that is, around topics such as the definitions of nanotechnology, nanogenomics, and human enhancement, how nanogenomics is being used now and might be used in the future, and varying viewpoints on the benefits and risks associated with nanogenomics. At the end were links and references to additional information. In Study 2 we used prompts to encourage different approaches to engaging with the background document, including one condition that asked students to organize the material they were reading into different perspectives that vary in the extent to which they see nanotechnology as risky versus beneficial. Because our initial analyses suggested that the information organization condition did not seem to impact student learning or other outcomes, in Studies 3 and 4, we experimentally varied the organization of the document itself, rather than prompting students to organize the information.

In Studies 3 and 4, we expanded the information provided and created two different versions of the background document. Both versions of the background information began with the same overview providing some definitions and examples of nanotechnology, nanogenomics, and nanomedicine. However, one version then presented the information related to ethical, legal, and social issues in a format similar to and inspired by the National Issues Forum (NIF-format). The other version included the same information but presented it in a topical format such as was used in Study 2. In the NIF-format materials, we identified explicitly opposing perspectives (e.g., "human enhancement as forward progress" vs. "human enhancement as unnecessary risk") and listed the action implications, evidence to support, pros and cons (trade-offs), and opposing points to each perspective. In the topical format, we did not explicitly identify opposing perspectives but instead discussed relevant topics that may impact people's views on use of nanotechnologies for human enhancement purposes, such as "the costs of not pursuing available benefits," "changing social concepts," "right to autonomy," and "unforeseen, unpredictable, unacceptable risks." Study 4 materials were largely similar to Study 3 materials, except that we included additional information about the risks or drawbacks of nanotechnology development in response to student concerns that the materials were positively biased.

Finally, in Study 5 we used only the NIF-formatted document, but we altered that document to create a stronger and weaker version. In studies 3 and 4, we found relatively few effects of the NIF versus topically organized documents but found students in our critical thinking conditions (described in the next section) were more negative about the background information. The use of stronger and weaker background documents was intended to explore whether our critical thinking students were more attentive to quality of information or simply more critical. To create the weaker documents, we altered some of the content and wording to introduce bias toward nanotechnology and removed some of the references supporting certain statements in the document (see detailed Study 5 methods in the supplementary files for comparison of the two versions).

## 2.3.3  Prompts for Cognitive Engagement

Readings about nanotechnology (given to students in A2 and referenced in A3) were accompanied by prompts to encourage deeper processing of the information. As discussed in more detail in Chap. 3, and in PytlikZillig, Hutchens, Muhlberger, and

Tomkins (2017), prior research has found that students can engage with reading materials in a variety of ways with different effects. For example, people may learn more if they are engaged in deep rather than surface-level processing, and some have found that the manner in which students take notes and organize the information impacts learning (Dinsmore & Alexander, 2012; Robinson & Kiewra, 1995).

In Study 2 we included three types of prompts during A2. One type of prompt encouraged students to organize the information using matrix note-taking, a type of note-taking that emphasizes comparison and contrast and has been found to improve learning over outline format note-taking (Robinson & Kiewra, 1995). Another type of prompt asked students to practice and then apply critical thinking strategies (e.g., looking for bias and examining the quality of evidence that was available to support claims). Critical thinking prompts were hypothesized to be directly beneficial to the goals of deliberation, which emphasize weighing of evidence. In addition, it was expected that the prompts would induce deeper processing of the information which could have positive impacts on learning. The third type of prompt was designed as a control that would evoke engagement and require students to respond but would not necessarily evoke deep or strategic engagement. The control prompts, which we often reference as the "feedback" condition, asked students to provide feedback on the readings and list "insights, realizations, reactions, or new things that you learned as a result of reading the background document or exploring the additional resources in that document."

As noted above, in Studies 3–5, we dropped the information organization (matrix note-taking) condition and focused only on the critical thinking and control prompts. In part this decision was made because the information organization condition appeared not to have much impact when compared to the control condition. In addition, beyond the classroom it seemed more feasible that we might be able to prompt people to think critically than to take notes in a certain way. At the end of Study 2, students were invited to engage in a fifth assignment during which we piloted some refinements to our prompts and administered additional personality measures. In Studies 3–5, we used refined critical thinking prompts in A2 that more gently nudged students to think critically, without asking them to practice thinking critically as was the procedure in Study 2. This change was implemented because our measures of cognitive-affective engagement suggested our Study 2 critical thinking prompts resulted in student *dis*engagement relative to the control prompts (PytlikZillig et al., 2017).

In addition, during Study 3, discussion facilitators who were leading groups of students in the critical thinking condition used discussion prompts that asked students to judge the quality of information shared and to be alert for sources of bias. In Study 4, we extended the critical thinking prompts to become part of the deliberative materials that all students received during A3. In Study 4 the critical thinking A3 prompts asked students to say what they thought about the scenario and included instructions reminding them of critical thinking skills that they could apply (e.g., citing sources and looking out for bias) as well as explicitly asking them to consider what someone with a different perspective might think about the issues. In the Study 4 A3 control condition, students were simply asked to write down their reactions to the scenarios, and no scaffolding of responses was provided. In addition, instead of explicitly asking what someone of a different perspective might think, a follow-up question simply asked what other questions they had about the topics under discussion. In Study 5, however, the critical thinking prompts were not used during the A3 discussions, but instead active facilitators sought to promote brainstorming, analysis, and synthesis of information and different perspectives, while passive facilitators merely read the discussion instructions and scenarios to the group.

### 2.3.4  Peer Discussion

In each of our Studies 2–4, we varied the presence versus absence of peer discussion during the in-class deliberative activities. We randomly assigned each student to discussion or non-discussion conditions. Students did not know whether they would be working alone or discussing issues with peers until immediately prior to A3. Students assigned to the discussion condition were then grouped with others who were in the same A2 condition (e.g., NIF vs. traditional, accompanied by critical thinking vs. control prompts). Students assigned to deliberate without peer discussion were directed to a separate room to read through and work on their deliberative activities during the same class period as the discussions took place. Both the discussion and non-discussion classrooms were monitored by a researcher or recitation instructor. All students, whether in the discussion condition or deliberating alone, had condition-appropriate background materials available to them either in hard copy form or via an online link. All students responded to the scenarios using online survey forms. A few copies of the survey forms were also available on paper in case of technical difficulties.

In general, more students enroll in the introductory biology course in the fall than in the spring, resulting in a larger sample sizes for those semesters. In fall of 2011 (Study 3), we thus assigned a convenient subset of the students to complete their deliberation activities via asynchronous online discussion with their group members. Because random assignment was not used to place these students in their condition, their data was treated as pilot data for exploring the impacts of students discussing online instead of face to face.

In fall of 2012 (Study 5), all students deliberated in groups. In part the decision to have all students in groups was due to student reports of much greater enjoyment and engagement when deliberating with their peers rather than alone. In addition, the choice to have students in groups allowed us to vary a specific theory-relevant aspect of discussion: attitudinal homogeneity, which is discussed in greater detail in Chap. 4. Positive and negative attitudes toward nanotechnology were determined by examining student responses to attitude questions answered at the end of A2. Once students were identified as tending to have relatively positive or negative attitudes, two-thirds were randomly assigned to homogeneous (positive or negative) groups and one-third to mixed (heterogeneous) attitude groups.

### 2.3.5   Active Facilitation During Discussion

In Studies 2–4, facilitators were used to lead group discussions during A3 due to recommendations made by many deliberative theorists and practitioners (Dillard, 2013). The facilitators were students recruited from prior semesters of the course who had previously taken part in the deliberative activities. The facilitators were trained by project researchers and given facilitation guides to ensure the use of common methods, prompts, and follow-up questions. In Studies 3–4, the facilitators were also given prompts specific to the critical thinking or control conditions to build on the prompts used in A2. In Study 5 active facilitators were instead given a list of prompts that might support and encourage brainstorming ideas, analyzing and evaluating different perspectives, and synthesizing information. In Study 5, facilitators were trained in both active and passive methods of facilitation in order to investigate the importance of active facilitation. Passive facilitators were instructed only to read the scenarios and questions used in the deliberation. Active facilitators were instructed to use the full range of facilitation techniques and the prewritten prompts and follow-up questions to evoke student interaction and consideration of different viewpoints.

## 2.4   For What Deliberative Engagement Outcomes?

Examination of the purported possible benefits and drawbacks of deliberation led us to focus upon, operationalize, and measure *outcomes* (which could serve as dependent variables) that included knowledge gains, changes in attitudes toward the topic of discussion, development of democratic or deliberative attitudes and other civic capacities such as political motivation and self-efficacy, changes in trust in scientists and regulators, and acceptance of policy resulting from such engagements. Table 2.6 provides a list of the major constructs measured across each of our studies and the assignments during which the measures were administered. Here, we give an overview of the outcome measures available for each study, but readers are referred to the supplemental materials for the specifics, including all items and how item wording may have been revised over time. In addition, here we focus on major measures used in at least two of the four reported studies. Additional measures and items are discussed in the supplemental materials (the detailed methods and materials accompanying this book).

### 2.4.1   Knowledge

Given the theoretical and empirical differences between subjective and objective knowledge (as discussed in greater detail in Chap. 3), in all studies, we assessed both subjective and objective knowledge with multiple items and nearly always at

**Table 2.6** Major constructs assessed and measures administered in each study

| | Study 2 | Study 3 | Study 4 | Study 5 |
|---|---|---|---|---|
| **Outcome measures** | | | | |
| Knowledge | | | | |
| Objective knowledge (T/F and mult. choice) | _2x__4 | 12x2_4 | 12x2_4 | 1_x2_4 |
| Confidence in objective knowledge (rate) | _____ | _2x2_4 | 12x2_4 | _____ |
| Subjective knowledge (rate) | _2x___ | 12x2__ | 12x2_4 | 1_x2_4 |
| Subjective learning (rate) | _____4 | __x2_4 | __x2_4 | __x234 |
| Attitudes toward nanotechnology | | | | |
| Risk-benefit (rate) | 12x234 | 12x234 | 12x234 | 1_x234 |
| Regulation-deregulation (rate) | _2x234 | 12x234 | 12x234 | 1_x234 |
| Values of nanotechnology (rating and/or percent) | 1___4 | 1___4 | 1___4 | ____4 |
| Opinions about nanotechnology (rate) | 1___4 | 1___4 | 1___4 | ____4 |
| Specific risk/benefits (likely x importance ratings) | 1___4 | 1___4 | 1___4 | ____4 |
| Change mind (predict, rate, and/or explanation (OE)) | _____4 | 1_x2_4 | 1_x234 | 1_x_34 |
| Initial/final views about nanotechnology (OE) | __x2_4 | 1___4 | 1___4 | 1*__4 |
| Perceptions of actors | | | | |
| Knowledge of and experience with actors (rate) | 1*___4* | 1___4 | 1___4 | ____4 |
| Trustworthiness of nanoscientists (rate) | 1*___4* | 1___4 | 1___4 | ____4 |
| Trustworthiness of policymakers (rate) | 1*___4* | 1___4 | 1___4 | ____4 |
| Policy scenario | | | | |
| Policy acceptance/support (rate) | ____4 | ____4 | ____4 | ____4 |
| Policy preference (rate) | ____4 | ____4 | ____4 | ____4 |
| Perceptions of public engagement procedures (rate) | ____4 | ____4 | ____4 | ____4 |
| Motivational variables | | | | |
| Deliberative citizenship | 1___4 | 1___4 | 1___4 | 1___4 |
| Political efficacy | 1___4 | 1___4 | 1___4 | 1___4 |
| Political motivation (intrinsic, introjected, extrinsic) | 1___4 | 1___4 | 1___4 | 1___4 |
| Perceived value of public engagement | | | | |
| Value of this ELSI engagement module | ____4 | ____4 | ____4 | ____4 |
| Overall value of ELSI engagements in college science (OE) | _____ | 1___4 | 1___4 | 1*__4 |
| Value of public engagement in general | _____ | _____ | _____ | ____4 |
| **Process measures** | | | | |
| Cognitive/affective engagement (rate) | __x23_ | __x23_ | __x23_ | __x23_ |
| Factors impacting attitudes/opinions (rate, some OE) | ____4 | ____4 | ____4 | __x234 |
| Group processes (individual ratings) | _____ | _____ | ___3_ | ___3_ |
| Group processes (facilitator ratings) | ___3_ | ___3_ | ___3_ | ___3_ |
| Perceived quality of background reading (rate) | __x2*_ | __x2_ | __x2__ | __x2__ |

(continued)

**Table 2.6**  (continued)

| | | | | |
|---|---|---|---|---|
| Perceived quality of assignment or module (rate) | __x234 | __x234 | __x234 | __x234 |
| Written questions and opinions (e.g., about scenarios) (OE) | ____3_ | ____3_ | ____3_ | ____3_ |
| Data quality checks | __x2_4 | __x2_4 | __x2_4 | __x2_4 |
| **Participant characteristics** | | | | |
| Demographics | 1____ | 1____ | 1____ | 1____ |
| Political party and ideology | 1____ | 1____ | 1____ | 1____ |
| Interest in politics | 1____4 | 1____ | 1____ | 1____ |
| Trust in institutions generally | 1____ | ____ | ____ | ____ |
| Dispositional trust (GSS and/or IPIP) | 1____ | 1____ | 1____ | 1____ |
| Subset of other Big 5 traits (IPIP) | 5 (NEOAC) | 1 (O only) | 1 (O only) | 1 (O and C) |
| Need for cognition | 1____ | 1____ | 1____ | 1____ |
| Right-wing authoritarianism | 1____ | 1____ | ____ | ____ |
| Cultural cognition | ____ | ____ | ____ | 1____ |

Note: OE indicates the measure is an open-ended text response. Numbers indicate whether the measure was administered during A1, A2, A3, and A4, respectively. A "2" preceding or following an x ("2x" or "x2") indicates the measure occurred during A2 before (preceding) or after (following) the reading of the background information accompanied by prompts. *Stars indicate instances where measures were administered but data may be limited by random assignment of items to subsets of students. When assessing the Big 5 traits using the International Personality Item Pool (IPIP), all Big 5 traits (neuroticism (N), extraversion (E), openness (O), agreeableness (A), and conscientiousness (C)) were assessed in Study 2, but the indicated subsets were assessed in Studies 3–5

multiple time points. **Subjective knowledge** was assessed with items such as "How familiar are you with nanotechnology?" and "How familiar are you with how nanotechnology is used in genetics research and development?" followed by five-point response scales ranging from "not at all familiar" to "extremely familiar." When multiple items were used, Cronbach alphas for scales created by averaging across subjective knowledge items at a specific time point were >0.80.

**Objective knowledge** was assessed with multiple choice and true-false items based on information that was presented in the A2 reading. In Studies 3 and 4, some of the objective knowledge items were accompanied by **confidence rating** ("I am _____ confident in my answer" with response options ranging from "not at all" to "completely"). As noted in Chap. 3, we evaluated the adequacy of our knowledge questions by examining the patterns of responses exhibited by individuals over time (Chatterji, 2003). Items sensitive to changes in knowledge should be more likely answered incorrectly prior to information presentation and relatively more likely to be answered correctly after information presentation. For each question, we examined the proportion of students who answered the question incorrectly prior to the reading and correctly after the reading and compared that percentage to the percent of students showing different patterns. For example, large percentages of students answering the question wrongly before and after information presentation would indicate an item was perhaps too difficult to detect knowledge changes or might indicate information was not adequately covered during information presentation and consideration. By examining the questions in this way, we were able to revise our knowledge questions over the first couple of studies. During each specific study, we chose items most likely to be sensitive to actual changes in student knowledge.

We also assessed student perceptions of their **subjective learning** during some of the assignments or for the module as a whole by asking questions such as "How much do you feel you learned about nanotechnology as a result of working on this assignment?" accompanied by a scale of five responses ranging from "nothing that I didn't know before" to "a great deal."

## 2.4.2  Attitudes Toward Nanotechnology

Attitudes toward nanotechnology were assessed in a number of ways, with certain questions asked during every assignment. As noted in Chap. 4, two rating questions were used over and over throughout the studies. One item asked about relative **risks and benefits**: e.g., "Based on what you know right now, do you think the risks of nano-biological research and development outweigh the benefits? Alternatively, do you think that the benefits outweigh the risks?" The risk benefits items were followed by a multipoint scale ranging from "The risks greatly outweigh the benefits" to "The benefits greatly outweigh the risks." Risk versus benefit items have been commonly used in prior research on public attitudes toward nanotechnology (e.g., see review by Satterfield, Kandlikar, Beaudrie, Conti, and Harthorn, 2009). The second item asked students about their perceptions of the **need to regulate** nanotechnology development. This item was often accompanied by a 100-point scale and read: "In your opinion, how much regulation is needed with respect to nano-biological research and development? Move the slider to reflect your view. 0 means that you believe there should be NO regulation of nano-biological developments. 100 means that you believe EVERY ASPECT of nano-biological research and development should be HIGHLY regulated."

Beyond those two single-item measures, which are analyzed in detail in Chap. 4, we also assessed a number of other more specific attitudes. These measures were often assessed at pre and post (e.g., A1 or A2 and again in A4) but not as frequently assessed throughout the assignments. A number of **opinion** items were obtained or adapted from prior research (Hamlett & Cobb, 2006; Scheufele & Lewenstein, 2005). Examples of these items include "The government will effectively manage any risks associated with nanotechnology" and "There are serious ethical problems associated with *not* quickly developing nanotechnologies." These items were typically accompanied by seven-point scale response scales ranging from "strongly disagree" to "strongly agree."

To assess different **values** associated with nanotechnology, participants also were presented with goals related to nanotechnology development and asked to assess their importance. Examples of these goals included "minimizing potential environmental risks" and "maximizing potential benefits for human enhancement." Response options typically fell on a five-point scale ranging from "unimportant" to "extremely important." In some studies, however, we also asked students to allot a certain percentage of total funding available to the different goals by explaining as follows: "There are, of course, only a limited number of resources (time, effort, money) that can be spent on each of the things that you rated above. So, please consider what percentage of the U.S. resources (i.e., total time, effort, and money) should be spent on each. Make sure that your percentages sum to 100%, and make sure that you give the items you rated the most important the greatest percentage of resources. What percentage of resources would you assign to each category?"

A number of more **specific risks and potential risks and benefits** were assessed using items taken or adapted from prior research on public attitudes toward nanotechnology (Cobb & Macoubrie, 2004; Lee, Scheufele, & Lewenstein, 2005). Examples of potential risks/benefits included nanotechnology leading to "The extension of human life expectancy" and "Pollution of the environment by nanomaterials." Responses fell on two different scales. The first, concerning likelihood of the risks or benefits, was typically a five-point scale ranging from "not at all likely" to "extremely likely." The second, concerning importance, was often in the form of a seven-point bipolar scale ranging from "very important to AVOID" to "very important to ACHIEVE," with "NOT IMPORTANT to avoid or achieve" representing the midpoint.

In addition to assessing attitudes directly, we also asked students to report their subjective sense of being likely to change or of having **changed their minds**. During the later assignments, after certain of the attitude questions, students were asked closed-ended rating questions such as, "Overall, from the time you began these exercises, until now, to what extent did you change your mind about the…question above?" followed by a five-point scale ranging from "not at all" to "a great deal." Similarly, students were sometimes asked whether they changed their opinions about nanoscientists or policymakers. In Study 4 we also asked students to rate the extent to which they *strengthened* their original opinions. Beginning in Study 3, during A1, we also asked how likely students thought it was that they would change their views. In some of the studies, the rating questions are accompanied by open-ended questions asking students to explain how and why their opinions changed or might change.

Finally, for each of our studies, we asked students to answer **open-ended opinion questions** regarding their attitudes toward nanotechnology. In most studies students were asked to answer similar or the same questions at least two times during the study, during A1 or A2 and again at A4. Overarching open-ended questions typically asked students to describe the developments relating to nanotechnology that should be prioritized or avoided as well as the regulations they felt should or should not be imposed. The students were also asked to give reasons for their views and at times were explicitly asked to consider opposing views and how they would answer those opposing views.

### 2.4.3   Perceptions of Actors: Nanoscientists and Policymakers

We assessed perceptions of nanoscientists and policymakers in most of our studies, and often at two different time points. Questions included items pertaining to **familiarity/certainty** (e.g., "How much do you know about nanoscientists?" and "To what extent do you feel certain in your views about policymakers who regulate nanotechnology?"). We also assessed confidence in and **perceptions of trustworthiness** of nanoscientists and policymakers. For example, students were asked, "To what extent do you have confidence in nanoscientists to…" "do their jobs well," "meet their professional responsibilities," and "make decisions about the most important directions for future development [of nanotechnology]." Responses typically fell on a five-point scale ranging from "not at all" to "a great deal." Following prior research on trustworthiness (Mayer, Davis, & Schoorman, 1995; PytlikZillig et al., 2016), we also often included items assessing specific trustworthiness and distrustworthiness components including whether participants feel nanoscientists and policymakers "are fair" or "…are primarily motivated by what will benefit them personally," "…don't really care about the long-term risks of their decisions," and "…are dishonest."

## 2.4.4  Policy Scenario: Policy Preference, Acceptance, and Support

At the end of the study during A4, we assessed policy preferences and acceptance/support. As described in Chap. 5, we conceptualized *policy acceptance/support* as a willingness to accept, tolerate, and not resist a policy, even if only with reservations or for a time. In contrast, we conceptualized *policy preference* as reflecting one's personally preferred policies: the extent to which people agree with, feel good about, and prefer a given policy.

We assessed **policy preference** by asking people to indicate the extent to which they were for or against a given policy. For example, participants were asked questions such as: "If legislation were being considered that would speed up nanogenomics research and development in the area of human enhancement by increasing funding and decreasing restrictions... Would you be FOR or AGAINST such legislation?" followed by a seven-point scale ranging from "strongly AGAINST" to "strongly FOR." In Studies 2–4 we randomly assigned students one of the two versions of the question, one asking about speeding up research and development and one asking about slowing it down. This was in order to find out if slowing down and speeding up policies were direct opposites or somehow different.

We assessed **policy acceptance** using an imagined scenario. In our study we were interested in whether salient public input processes could increase public acceptance, and thus we wrote a scenario that was designed to first make those processes salient and then assess acceptance/support in light of those salient processes. In our scenario, the government purportedly listened to input from the public, and input was purportedly gathered according to the same procedures the student had experienced. Then the government made a decision that was consistent with the public input. In Studies 2–4, after describing the scenario we had designed, we assessd policy acceptance by asking participants if they agreed "The government made the right choices with regard to this issue," "The government made the same choices you would have made," and/or if "The government should have made different decisions about the issue" (reversed). In Study 5 we directly assessed **policy**

**support/resistance** in addition to acceptance by asking if participants agreed or disagreed that "I would support this decision made by the government [in the scenario just read]," and "I would resist this decision made by the government" (reversed). In Study 5 we also included an acceptance item: "Because of the processes used, I would accept this decision made by the government."

Regarding the validity of assessing policy acceptance or support as separate from policy preferences, students were allowed to explain their answers, and some of their explanations did indicate that at least some students perceived the questions as different from one another. Specifically, the acceptance answers by at least some students took into account the results of the public engagement processes used to come to the decision. As an example, here is an explanation given by a student in Study 4 who also indicated a preference for pro-nanotechnology policies (expressing that he was "strongly against" slowing down research and increasing regulation). Despite the pro-development preference this student indicated, he also agreed (not strongly, but also not slightly) that the government in the scenario made the same choices that he would have made by slowing down development of nanotechnology and increasing regulations. As an explanation he wrote:

> If the majority of people feel that they're not ready for the technology then it's hard to change that. The government is supposedly the will of the people...If people of a nation-state [choose] not to participate in new technologies then that is the will of the country, the government should abide by that will....

### 2.4.5  Motivational Variables

We also assessed certain individual differences in motivation and attitudes or beliefs that, based on prior theory, might be expected to be affected by participation in deliberative engagements. In each study, at both A1 and A4, we assessed attitudes toward deliberative engagements or **deliberative citizenship** using five items relating to the value of deliberation as an element of active citizenship. These included the statements "A good citizen should listen to people who disagree with them politically" and "A good citizen should be willing to justify their political views." These items were taken from Muhlberger and Weber (2006), and when averaged into a scale, the scale internal consistency as assessed with Cronbach alpha was always >0.75 in each of the four studies.

Political self-efficacy can also be impacted by engaging in political discussions with others. We assessed **political self-efficacy** at A1 and A4 in each study, using four to six items and resulting in a scale with Cronbach alphas >0.70 for each of the studies. Examples of these items include the statements "Sometimes politics and government seem so complicated that a person like me can't really understand what's going on" which was reverse-scored, and "I consider myself well-qualified to participate in politics." These statements were adapted from similar items used in the American National Election Study (2000).

Finally, drawing from self-determination theory which suggests that people's motivations can become internalized over time and with various experiences, we assessed three types of **political motivation** in both A1 and A4 in each study: intrinsic, introjected, and extrinsic. The items used for these measures were adapted from Koestner, Losier, Vallerand, and Carducci (1996) and Losier and Koestner (1999). To measure the extent to which a student felt motivated to be politically engaged by factors that are *intrinsic* to their concept of self, we used statements such as "I follow political and social issues because I want to learn more things" and "I follow political and social issues because I think it's important." The extent to which a student felt motivated to be politically engaged by factors they had internalized from outside sources (*introjected motivation*) was assessed with items such as "I follow political and social issues because it bothers me when I don't." The extent to which a student felt motivated to be politically engaged by external factors (*extrinsic motivation*) was assessed with one item such as "I follow political and social issues because that's what I'm expected to do." When multiple item scales were used to assess these constructs, the Cronbach alphas were >0.70.

### 2.4.6  Evaluation of Public Engagement

We assessed student perceptions of public engagement and its usefulness in various ways. In our **module evaluation** questions, we often asked "Overall, to what extent was the public participation module a beneficial part of your learning in this course?" with responses to both on a five-point scale ranging from "not at all" to "a great deal." In addition, to better understand perceptions of conducting **ELSI engagement activities in the context of science courses**, we asked students at different points, sometimes both before and after engaging in all of the module activities, to answer an open-ended question that read, "In your opinion, how important is it that science students--including beginning science students such as you and your classmates--learn how to think about the ethical, legal and social issues (ELSI) pertaining to science? In 2–3 sentences, give your answer and a brief explanation of why you think as you do." Finally, to assess perceptions of the **value of public engagement in general**, in A4 of Study 5, we additionally asked "Some people feel that the government should primarily rely on expert opinions, not citizen opinions when making policy decisions. What do you think? In your opinion, how much weight should government give to citizen opinions (like you wrote above) when making decisions about the future of nanotechnological development and regulation?" This question was accompanied by a five-point response scale ranging from "None! The government shouldn't be considering opinions of everyday people like me" to "A lot of weight! The government should take opinions like mine very seriously."

## 2.5  How and Why: Mediators and Moderators

Having identified *features* and *outcomes*, the next challenge is to connect them via explanatory theories. What mechanisms might explain why certain public engagement features might or might not connect to certain outcomes in certain contexts? And what moderators might impact different effects? One primary mechanism examined in our studies was "engagement"—or, rather, the varieties of ways that people can engage (PytlikZillig et al., 2013). Other process measures assessed group processes that took place in A3; student perceptions of the quality of the background readings, assignments, and public engagement module as a whole; whether and why (or why not) students felt they changed their mind on different opinion or attitude questions; and their open-ended responses to questions about the development and regulation of nanotechnology. Potential moderators that we assessed included participant characteristics such as their demographics, their political ideology and party, and various personality traits such as openness, conscientiousness, need for cognition, and dispositional trust.

### 2.5.1   Cognitive-Affective and Behavioral Engagement

It is interesting that although engagement is, in name, central to public engagement, engagement research does not often focus on the varieties of ways individuals might be engaging. How do people feel when they are engaging? Are they bored, interested, and annoyed? Maybe they are distracted by things happening on their Instagram feed. If they are fully engaged with the topic, are their minds open and listening to another person's perspective they have not heard before, or are their minds busy plotting their rebuttals to earlier remarks?

As described elsewhere (PytlikZillig et al., 2013), engagement is a varied and multifaceted state that includes the affective, cognitive, and behavioral experiences of individual participants at different points during the engagement activities. We drew from a number of theories in developing self-report engagement scales, including cognitive theories of deep and surface cognitive processing (Chin & Brown, 2000; Dinsmore & Alexander, 2012), and educational theories of metacognition and active learning strategies (McCormick, 2003; Veenman, Van Hout-Wolters, & Afflerbach, 2006; Vermunt & Vermetten, 2004). Drawing from theories of emotion and affect, we incorporated items asking about anger and boredom (Fahlman, Mercer-Lynn, Flora, & Eastwood, 2013; Harmon-Jones, Schmeichel, Mennitt, & Harmon-Jones, 2011). Drawing from personality and social psychology, we devised measures of states of open-mindedness, creativity, closed-mindedness, conscientiousness, and social engagement (Akbari Chermahini & Hommel, 2012; Fleeson, 2001; PytlikZillig, 2001). Then, in our studies, we asked participants to self-report how they had engaged during different activities—such as reading and responding to background readings in A1 or deliberating alone or with peers in A3. This allowed for the investigation of the roles of certain engagement "states" that are explicitly or implicitly referenced by numerous theories that could be applied in the contexts of public engagements.

To assess engagement, participants were asked to assess various statements, each beginning with the stem "During the assignment, I…" (e.g., "felt focused"). Responses fell on a five-point scale ranging from "not at all" to "a great deal." Items were taken from/reported in PytlikZillig et al. (2013) and were intended to assess each of the eight different ways of engaging: active learning (e.g., "identified questions that I still had about the topics"), conscientious (e.g., "gave careful consideration to all of the options presented"), open-minded (e.g., "felt open to hearing new ideas about the topics"), social (e.g., "discussed my ideas about the topics with others"), creative (e.g., "used my imagination"), disinterested (e.g., was impatient to get this over"), angry (e.g., "felt angry"), and closed-minded (e.g., "felt like my mind was already made up"). Finally, in addition to the items assessing states of engagement experience, we also at times asked students to self-report the amount of time that they spent on various portions of assignments.

## 2.5.2  Self-Reports of Influences on Attitudes

In addition to the questions about whether and how much attitudes changed (variables listed under outcomes), in some of the studies, we asked students to rate **influences on their attitudes**, at times including whether and how much a specific assignment influenced their specific opinions or attitudes about nanotechnology. In some studies, they rated how much certain factors (e.g., talking about the issue with others, background reading, views of people important to them) impacted the views that they expressed in the surveys. At times the rating questions were accompanied by open-ended questions asking students to elaborate on why they changed their views.

## 2.5.3  Participant and Facilitator Perceptions of Group-Relevant Processes

Another set of process measures, used in Studies 4 and 5, asked student participants involved in discussions to reflect on their discussion experiences. Questions typically focused on perceptions of their group and group members and perceptions of their facilitator. In addition, some of our studies include ratings made by the facilitators of the group processes.

Related to participant **perceptions of their group and group members**, some statements were relevant to identification with the group. Examples of these items include "I identified with my group during the discussion" and "I felt like an 'outside' member of my group," reverse coded. Responses were on a seven-point scale ranging from "strongly disagree" to "strongly agree." Other questions measured group consensus such as "How much disagreement was there within your group at the beginning of discussion?" and "How much agreement was there within your group at the end of discussion?" Additional questions included "Overall, how do you feel about the other people in your group?" (seven-point scale from very negative to very positive) and "How satisfied were you with your small group discussion in class today?" (five-point scale from not at all satisfied to extremely satisfied). Some questions focused on individual group members. Participants were asked to identify each member of their group by first name and last initial. They were then asked questions referencing *each* individual. For example, "Before today's discussion, how familiar would you say you were with _____?" (with responses falling on a five-point scale ranging from "not at all familiar" to "extremely familiar") and "Regarding the ethical scenarios you just discussed, how similar do you think _____'s opinions are to your opinions?" Response options were comprised of five-point scales ranging from "not at all similar" to "extremely similar."

Related to **perceptions of their facilitator**, participants in Study 5 were asked to indicate the extent of their agreement with statements regarding how active or passive their facilitator was. These statements all began with the phrase "The moderator of our group…." Examples of these items include "was very active in leading our group discussion" and "summarized what he/she heard the group saying." Responses fell on a seven-point scale ranging from "strongly disagree" to "strongly agree." These questions served as a check for our manipulation of facilitator activity.

In each of the studies, **facilitator perceptions** of group processes were assessed at the group level, with the questions being refined with each study. For example in Study 2, facilitators gave their impressions regarding how interested participants were in the discussion questions and how much the students stayed on topic and discussed alternative perspectives. In Study 3 facilitators additionally gave impressions about whether students used critical thinking skills, such as providing and evaluating evidence for different perspectives. In Study 4, facilitators also reported some of the topics that came up in discussion, and in Study 5, they were asked to perform a self-check of how well they adhered to the active and passive conditions during their facilitation.

### 2.5.4   Assignment and Information Evaluations

After the most substantive of the assignments (A2 reading, A3 deliberation) and at times at the end of the entire set of public engagement activities (during A4), students were asked to evaluate the quality of various components of the public engagement. For example, they rated the **quality of the background information** by responding to statements beginning with the stem "The information provided in this assignment was…" and ending with phrases such as "unbalanced" and "fair in its presentation of the issues," to assess perceptions of bias. They also rated the clarity or understandability of the background information and its quality in terms of accuracy and thoroughness. In addition to rating the background information, within specific studies, students were at times also asked to rate the quality of A2 as a whole or the usefulness and value of the entire public engagement "module" (set of activities) used in the course.

### 2.5.5   Written Reponses and Comments

Open-ended responses also were gathered from our students. In addition to the ones already described in previous sections of this chapter, student written responses to open-ended questions during the A3 deliberation are also available for merging with our data sets by request. The open-ended data include responses to the scenarios that they read in Studies 2–4 and a drafted and revised "law" for the regulation of nanotechnology that students were asked to write during A3 of Study 5. Often students also left open-ended comments at the end of a given assignment or study.

### 2.5.6 Data Quality Checks

For the most part, our open-access data sets include all consented data regardless of quality. Exceptions are noted in our supplemental materials. In each of the studies, we included some assessments of data quality, including questions which overtly asked students if they completed the questions honestly (vs. randomly, without reading, or felt they should answer in a certain way that differed from their honest response). In Study 5 we also included, in some assignments, items explicitly asking participants to choose a certain response (e.g., directing them to choose "strongly agree"). Students failing to follow those instructions may have been answering the survey inattentively. In addition, it may be possible to ascertain random or inattentive responding by examining the pattern of responses to certain of our measures—such as whether students classified presumed potential negative outcomes as negative (e.g., pollution).

### 2.5.7 Demographics and Individual Differences

Other potential moderators assessed in our studies include measures of demographics and individual differences. In all studies, we assessed self-reports of **demographics** including age, gender, year in school, typical grades, and prior experience with ethics coursework. In Studies 2–4 we also asked for self-reports of both parents' highest level of education.

In all four studies, we also included multi-item measures of ideological identity, interest in politics, dispositional trust, and need for cognition. In some but not all studies, we also assessed trust in institutions, cultural cognition, authoritarianism, and certain of the Big 5 traits of openness, agreeableness, emotional stability, extraversion, and conscientiousness. The details of these additional measures are in our supplemental materials.

## 2.6 Conclusion

In this chapter, we presented a large number of comparisons and details of variations between studies in order to give readers an overview that will allow them to assess whether our data may be useful for their own purposes, as well as to provide background on the measures we reference in our remaining chapters. Readers will find even more detail in the documentation that accompanies the data sets in the supplemental materials.

Next, to provide exemplars of how our data may be used to test various theories, we turn our attention to a much narrower set of variables that comprise some of the most desired outcomes of public engagement: increases in knowledge, changes in attitudes or opinions, and acceptance of policy decisions.

# References

Akbari Chermahini, S., & Hommel, B. (2012). Creative mood swings: Divergent and convergent thinking affect mood in opposite ways. *Psychological Research, 76*, 634–640.

Chatterji, M. (2003). *Designing and using tools for educational assessment*. Boston, MA: Allyn & Bacon.

Chin, C., & Brown, D. E. (2000). Learning in science: A comparison of deep and surface approaches. *Journal of Research in Science Teaching, 37*, 109–138.

Cobb, M. D., & Macoubrie, J. (2004). Public perceptions about nanotechnology: Risks, benefits and trust. *Journal of Nanoparticle Research, 6*, 395–405.

Delgado, A., Kjølberg, K. L., & Wickson, F. (2011). Public engagement coming of age: From theory to practice in STS encounters with nanotechnology. *Public Understanding of Science, 20*(6), 826–845.

Dillard, K. N. (2013). Envisioning the role of facilitation in public deliberation. *Journal of Applied Communication Research, 41*(3), 217–235.

Dinsmore, D. L., & Alexander, P. A. (2012). A critical discussion of deep and surface processing: What it means, how it is measured, the role of context, and model specification. *Educational Psychology Review, 24*, 499–567.

Fahlman, S. A., Mercer-Lynn, K. B., Flora, D. B., & Eastwood, J. D. (2013). Development and validation of the multidimensional state boredom scale. *Assessment, 20*, 68–85.

Fleeson, W. (2001). Toward a structure- and process-integrated view of personality: Traits as density distributions of states. *Journal of Personality and Social Psychology, 80*, 1011–1027.

Hamlett, P. W., & Cobb, M. D. (2006). Potential solutions to public deliberation problems: Structured deliberations and polarization cascades. *Policy Studies Journal, 34*(4), 629–648.

Hamlett, P. W., Cobb, M. D., & Guston, D. H. (2008). National Citizens' Technology Forum: Nanotechnologies and Human Enhancement: Arizona State University. Retrieved from http://cns.asu.edu/sites/default/files/library_files/lib_hamlettcobb_0.pdf (CNS-ASU Report #R08–0003).

Harmon-Jones, C., Schmeichel, B. J., Mennitt, E., & Harmon-Jones, E. (2011). The expression of determination: Similarities between anger and approach-related positive emotion. *Journal of Personality and Social Psychology, 100*, 172–181. https://doi.org/10.1037/a0020966.

Koestner, R., Losier, G. F., Vallerand, R. J., & Carducci, D. (1996). Identified and introjected forms of political internalization: Extending self-determination theory. *Journal of Personality and Social Psychology, 70*(5), 1025–1036.

Labov, J. B., Reid, A. H., & Yamamoto, K. R. (2010). Integrated biology and undergraduate science education: A new biology education for the twenty-first century? *CBE Life Sciences Education, 9*(1), 10–16. https://doi.org/10.1187/cbe.09-12-0092. Retrieved from http://www.ncbi.nlm.nih.gov/pmc/articles/PMC2830155/.

Lee, C.-J., Scheufele, D. A., & Lewenstein, B. V. (2005). Public attitudes toward emerging technologies examining the interactive effects of cognitions and affect on public attitudes toward nanotechnology. *Science Communication, 27*(2), 240–267.

Losier, G. F., & Koestner, R. (1999). Intrinsic versus identified regulation in distinct political campaigns: The consequences of following politics for pleasure versus personal meaningfulness. *Personality and Social Psychology Bulletin, 25*(3), 287–298.

Mayer, R. C., Davis, J. H., & Schoorman, F. D. (1995). An integrative model of organizational trust. *Academy of Management Review, 20*, 709–734.

McAvoy, P., & Hess, D. (2013). Classroom deliberation in an era of political polarization. *Curriculum Inquiry, 43*(1), 14–47.

McCormick, C. B. (2003). Metacognition and learning. In W. M. Reynolds & G. E. Miller (Eds.), *Handbook of psychology volume 7: Educational psychology* (pp. 79-102). Hoboken, NJ: Wiley.

Muhlberger, P., & Weber, L. M. (2006). Lessons from the virtual Agora project: The effects of agency, identity, information, and deliberation on political knowledge. *Journal of Public Deliberation, 2*(1), 1–39.

National Research Council. (2009). *A new biology for the 21st century: Ensuring the United States leads the coming biology revolution.* Washington, DC: National Academy of Sciences. www.nap.edu/catalog.php?record_id=12764.

PytlikZillig, L. M. (2001). *Extraversion as a process: The effects of extraverted "states".* (Ph.D. Dissertation), University of Nebraska-Lincoln, Lincoln, Nebraska.

PytlikZillig, L. M., Hamm, J. A., Shockley, E., Herian, M. N., Neal, T. M., Kimbrough, C. D., ... Bornstein, B. H. (2016). The dimensionality of trust-relevant constructs in four institutional domains: Results from confirmatory factor analyses. *Journal of Trust Research.* https://doi.org/10.1080/21515581.2016.1151359.

PytlikZillig, L. M., Hutchens, M., Muhlberger, P., Wang, S., Harris, R., Neiman, J., & Tomkins, A. J. (2013). The Varieties of Individual Engagement (VIE) measure: A confirmatory factor analysis across two samples and contexts. *Journal of Public Deliberation, 9*(2), Article 8. Retrieved from http://www.publicdeliberation.net/jpd/vol9/iss2/art8.

PytlikZillig, L. M., Hutchens, M. J., Muhlberger, P., & Tomkins, A. J. (2017). Prompting deliberation about nanotechnology: Information, instruction, and discussion effects on individual engagement and knowledge. *Journal of Public Deliberation, 13*(2), Article 2. Retrieved from https://www.publicdeliberation.net/jpd/vol13/iss2/art2/.

PytlikZillig, L. M., & Tomkins, A. J. (2011). Public engagement for informing science and technology policy: What do we know, what do we need to know, and how will we get there? *Review of Policy Research, 28,* 197–217.

Robinson, D. H., & Kiewra, K. A. (1995). Visual argument: Graphic organizers are superior to outlines in improving learning from text. *Journal of Educational Psychology, 87,* 455–467.

Royal Society/RAE. (2004). *Nanoscience and nanotechnologies: Opportunities and uncertainties.* London, UK: Royal Society and Royal Academy of Engineering. https://royalsociety.org/~/media/Royal_Society_Content/policy/publications/2004/9693.pdf.

Satterfield, T., Kandlikar, M., Beaudrie, C. E., Conti, J., & Harthorn, B. H. (2009). Anticipating the perceived risk of nanotechnologies. *Nature Nanotechnology, 4*(11), 752–758.

Scheufele, D. A., & Lewenstein, B. V. (2005). The public and nanotechnology: How citizens make sense of emerging technologies. *Journal of Nanoparticle Research, 7*(6), 659–667.

Veenman, M. V., Van Hout-Wolters, B. H., & Afflerbach, P. (2006). Metacognition and learning: Conceptual and methodological considerations. *Metacognition and learning, 1*(1), 3–14.

Vermunt, J., & Vermetten, Y. (2004). Patterns in student learning: Relationships between learning strategies, conceptions of learning, and learning orientations. *Educational Psychology Review, 16,* 359–384. https://doi.org/10.1007/s10648-004-0005-y.

# Chapter 3
# Knowledge

**Abstract** This chapter examines what is associated with increases in both objective and subjective knowledge about nanotechnology as the result of participating in a public engagement. The results are replicated and compared across three different public engagements, all using undergraduate students as participants. Knowledge is examined at four different time points, allowing researchers to understand when learning is most likely occurring. Results indicate that participants showed gains in knowledge over the course of the public engagement, with the biggest gains shown after reading the materials as compared to participating in the group discussions. The structure of the materials did not directly influence knowledge gain; however, there were indirect effects of encouraging critical thinking on knowledge via cognitive engagement. These results highlight the importance of cognitive engagement to understand when learning occurs, as well as some of the opportunities that may exists for remote deliberations, given the importance of the reading materials over the discussion.

**Keywords** Cognitive engagement · Mediation · Subjective knowledge · Objective knowledge · Critical thinking · Information organization

> Everyone is entitled to his own opinion, but not to his own facts. – Daniel Patrick Moynihan

A common expectation among engagement scholars is that participating in public engagements will lead to better-informed citizens (e.g., Gastil & Dillard, 1999; Selin et al., 2017). We want citizens to know about policies and technologies and then make decisions about them. We want citizens to have reactions that are informed by facts that experts can agree upon, rather than using "alternative facts." Recent research on online rumors (e.g., Garrett, Weeks, & Neo, 2016) and boomerang effects (e.g., Hart & Nisbet, 2012) (when persuasion attempts result in people adopting the opposite of what the persuasion was pointing toward) reinforces the

**Electronic supplementary material**: The online version of this chapter (https://doi.org/10.1007/978-3-319-78160-0_3) contains supplementary material, which is available to authorized users.

idea that simply being aware of a policy is not enough. This research suggests that rumors are prevalent in online contexts, and individuals will often cling to incorrect facts even after hearing that their information is incorrect. Knowledge is often at the forefront of discussions about the role that citizens should play in science policy discussions and was a crucial outcome examined in our own work. The research here deals with a few key questions: What situations will increase learning during public engagements? What features will help participants best understand the facts relevant to the policies and technologies under consideration? We address knowledge in our studies by assessing what the student engagement participants learned over time, and what helped or hindered that learning.

## 3.1  Why Does Knowledge Matter?

A disappointing but well-known fact is that Americans are especially ill-informed about a variety of political issues. Late-night comics make a joke of this in the common "man-on-the-street" interviews that highlight how little the average person knows as the audience laughs along. A 2014 video created by a group of Texas Tech students went viral when it illustrated fellow students' lack of knowledge of the name of their vice president and who won the Civil War.[1] While we may laugh at our own expense, what people know does, at least theoretically, have important implications for their reactions to policy proposals. Democratic theorists assert that individuals need to be knowledgeable in order for them to participate effectively in the public sphere. Delli Carpini and Keeter (1996, pp. 5) state that "knowledge provides the raw material that allows citizens to use their virtues, skills, and passions in a way that is connected meaningfully to the empirical world." Therefore, having knowledge is an important qualification in order to self-govern and to have meaningful debates about various issues.

Unfortunately, knowledge is not equally distributed among citizens, and how that can be ameliorated is the topic of decades of research in political science and communication. At the core of research on knowledge in political and policy-related contexts is the assumption that inequalities in knowledge will lead to inequalities in participation and ultimately inequalities in the benefits that can be obtained from participation (Dutwin, 2003). Thus, in addition to raising knowledge levels, it would be valuable to help people to reach more equal levels of knowledge through public engagement activities.

Deliberative democratic theorists assert that the best decisions will be made for the greatest number of people when citizens are informed about all sides of an issue. We can trace this line of thinking back to John Stewart Mill and the "Marketplace of Ideas." Essentially, Mill believed that, if we let the information be known, the best ideas will rise to the top (Mill, 1860). However, being aware of something is not the same as having knowledge. The current debate about fake news and the speed with which rumors can spread online highlight that difference (e.g., Allcott &

---

[1] https://www.youtube.com/watch?v=yRZZpk_9k8E

Gentzknow, 2017). Instead, the marketplace of ideas needs to inform us more substantially—we need to fully understand what is happening rather than simply being aware that a policy or debate over policy exists. The purpose of public engagements is to provide a variety of information and perspectives to individuals, with the hope that they leave more knowledgeable, and with clearer and more evidence- and fact-consistent beliefs (Delli Carpini, Cook, & Jacobs, 2004).

We need to note that knowledge is sometimes viewed as an inappropriate focus in studies of engagement, because it is often associated with so-called deficit models used in science communication (e.g., Sturgis & Allum, 2004). These models propose that citizens do not believe in or value various scientific findings or do not behave in a way that is desired, simply because they do not have adequate knowledge (or trust or empathy, etc.). In essence, deficit models assume an ignorant public and assert that everyone will agree once the problem of ignorance (or mistrust, etc.) is fixed. While there is some research to support this view (van der Linden, Leiserowitz, Feinberg, & Maibach, 2015), the vast majority of research paints a much more complex picture (Druckman & Bolsen, 2011; Sturgis & Allum, 2004). Some would argue that, even given the same set of facts, different individuals will come to different conclusions based on their own values and other considerations. Even those who critique the role of knowledge in producing other public engagement outcomes, however, do not question that knowledge is beneficial for citizen engagement. That is, while knowledge may not always lead to changes in behavior or attitudes, we would argue that knowing more would be an indication of a successful public engagement. In public engagement, deficit model thinking, or simply providing a set of facts that will remove ignorance and assure the "right" way of behaving, is not the focus. Discussion of pros and cons is often explicitly encouraged rather than asserting one way as preferred, and participants are encouraged to share their own perspectives.[2] However, engagement scholars believe that participants do need to be working from the same set of facts, which can be provided at several stages throughout the engagement activities (Goodin & Niemeyer, 2003; Muhlberger & Weber, 2006).

## 3.2   How Can Public Engagements Foster Increases in Knowledge?

Prior research on participation in public engagements have found that participants feel like they know more (Powell & Kleinman, 2008) and in many cases actually do know more facts (Fishkin, 1997; Luskin, O'Flynn, Fishkin, & Russell, 2014) about the topic at hand after participating. The research that has explicitly examined when individuals are most likely to learn suggest that most of the learning from participating in public engagements happens while individuals are reading the information provided prior to deliberation, and the discussion itself does not add much

---

[2] https://www.nifi.org/

additional information (Goodin & Niemeyer, 2003; Muhlberger & Weber, 2006). This research is the exception however, and most look at the impact of the entire public engagement experience. Much of the research and theory about everyday discussion would suggest that the discussion is essential in helping individuals crystallize what they learn from other sources (Eveland & Hively, 2009; Mansbridge, 1999; Sanders, 1997); however, the research supporting that fact in the presence of readings and discussion is rather scant.

There are a host of theories that can help explain why participating in public engagements would lead to an increase in knowledge, but they can generally be tied to two different perspectives. One perspective is simply that providing information will lead to an increase in knowledge, with the caveat that the information is attended to in some fashion. A wrinkle to this perspective in the public engagement literature focuses on *how* the material is being presented. For example, the National Issues Forum recommends that information be presented in a balanced, nonpartisan fashion. The second perspective instead focuses on cognitive engagement with various aspects of the materials as the reason behind why individuals learn. Put simply, participating in an engagement will motivate you to think (or cognitively engage) with the concepts somehow, which in turn will foster learning. Along these lines, there is evidence that when participants know their input will matter, this increases motivation, and they are more likely to cognitively engage with the materials (Powell & Kleinman, 2008). However, the reasoning behind the "presentation of materials" arguments is directly tied to presenting the material in a way that will foster deeper or more effective cognitive engagement, so the two perspectives are not completely distinct. Over our series of studies, we examine both of these perspectives to try to better understand how best to foster increases in learning.

### 3.2.1 Informational Presentation

Public engagements will typically give participants some sort of information prior to coming together as a group. During this preliminary information presentation, which may be in the form of materials to be read in advance of the deliberative gathering or an initial information session during the gathering, the "facts" are stated in advance of the deliberative portion of the engagement. There is generally extensive care put into the creation of these documents to try to provide the nuance of various perspectives from a balanced, nonpartisan perspective. This is done so individuals will come to the deliberative engagement prepared to knowledgeably engage in the various sides of an issue. There are theoretical reasons to believe that how this information is presented can influence how citizens respond to that information. In general, practitioners argue that at a minimum, the information needs to be presented in a balanced manner (Burkhalter, Gastil, & Kelshaw, 2002; Lukensmeyer & Torres, 2006). However, even while meeting that criterion, there are various ways in which that information might be structured.

As noted in Chap. 2, we focused on two different informational presentation strategies that are typically used in public engagements. The format that we expected would have a positive impact on learning was a *pro-con* organization style. Such a style is typically used, for example, in the National Issues Forums. This strategy explicitly compares the various perspectives on the information that is being presented. The other presentation strategy was a topical organization that does not explicitly call out the different perspectives. Our expectation was the pro-con organization would facilitate learning based on prior educational research that has examined different note-taking styles. Scholars find that making comparisons while taking notes improves learning over simple linear methods (Robinson & Kiewra, 1995). Furthermore, texts structured in a compare-contrast form tend to be associated with improved recall, compared to texts organized linearly or descriptively (Bohn-Gettler & Kendeou, 2014).

The second experimental manipulation relevant to knowledge, which also focused on the presentation of the materials, was the inclusion of prompts encouraging participants to think critically. The justification for these prompts encouraging learning is tied to research in education that shows that students learn differently depending on their goals (Bohn-Gettler & Kendeou, 2014). Participants were simply asked either to provide feedback or to critically evaluate the material at several points throughout the reading. By encouraging students to use critical thinking skills, we expected that individuals would pay closer attention to the material and process the information more deeply, thereby facilitating learning. How individuals process information—or how they engage with the materials cognitively—is an important consideration in many literatures. This was also true for our own studies, and we assessed cognitive engagement several ways throughout our series of experiments.

### 3.2.2 Cognitive Engagement

The theoretical reasoning behind why a pro-con information organization or critical thinking prompts would facilitate learning all goes back to how individuals may process the material with which we present them. In essence, we were testing whether or not we could get participants to engage with our materials in a deeper and more effortful fashion. We also then explicitly measured the extent to which participants said they were engaged while participating. The need for individuals to process the information they are presented with in order to learn is central to several theories across a variety of fields. Communication scholars will often invoke the communication mediation model (Eveland, 2001; Shah et al., in press) which indicates that the extent to which individuals elaborate, which is engaging in additional thinking, or engage in perspective taking after being presented with information will influence how much they learn. In psychology, dual process models of attitude change—which have also been applied to learning—indicate that longer-lasting effects in attitudes occur when individuals systematically process information (Eagly & Chaiken, 1993; Petty & Cacioppo, 1986). Systematic processing implies individuals are both motivated and able to expend cognitive effort processing information. The theories that we relied on most heavily when examining cognitive engagement are those that reside in education psychology. In particular, that research suggests that the greatest learning will occur when individuals use deep rather than surface-level processing strategies, which is more likely when they are motivated.

## 3.2.3  Forms of Cognitive Engagement

Research in areas such as cognitive and educational psychology find that one of the most important predictors of whether, how much, and how robustly people learn is *how* they engage with information. Much of this research has been conducted in educational contexts and examines a variety of different types of study strategies, categorizing ways of studying new materials as involving surface-level versus deep cognitive processing, or as being strategic and conscientious. For our experiments here, we created eight scales designed to measure various facets of individual engagement (for an in-depth examination of the construction and validation of those scales, please see PytlikZillig et al., 2013). The eight types of engagement that we examined were active learning, conscientious, uninterested, creative, open-minded, close-minded, angry, and social (measures used to assess these are discussed in Chap. 2). Active learning and conscientious and uninterested engagement relate to participants' motivations while engaging. *Active learning* engagement is assumed to occur when participants' acknowledge that they are trying to deeply process the information with which they are presented. This deep processing is likely to be tied to increased knowledge. *Conscientious* engagement refers to an individual's desire to be careful or thoughtful while examining information, which also would be associated with enhanced learning. *Uninterested* engagement is characterized by low motivation and boredom, which would likely impede learning. Creative, open-minded, and close-minded engagement focus more on how individuals are participating. *Creative* engagement focuses on whether individuals are attempting to "think outside the box" and potentially use multiple and unexpected perspectives when participating. While this could increase learning, it is also possible that creative engagement could have a negative impact on learning due to the possibility of distraction by irrelevant or incorrect information. *Open-minded* and *close-minded* engagement assess the extent to which individuals are willing to be open versus closed to other's opinions. *Angry* engagement examines participants' negative emotional engagement while participating, which may indicate that participants are engaging in defensive strategies and therefore are less likely to learn. *Social* engagement is designed to assess the extent to which a participant connected and interacted with others while participating.

### 3.2.4 Need for Cognition

The last element we examined was an individual difference variable that is frequently used when examining knowledge gain or learning—need for cognition (NFC). NFC is the general tendency to enjoy and use effortful cognitive processing strategies (Cacioppo et al., 1996). Persons with high NFC should be most likely to learn when participating in public engagements. Such people have been identified as especially likely to participate in deliberations, be more resistant to the arguments of others, and have more influence (Delli Carpini et al., 2004). In our case, we were more interested in various features that could increase knowledge across the board once we accounted for differences in NFC. Essentially, we wondered if we could level the cognitive playing field with our various informational presentation strategies.

## 3.3 What Do We Mean by Knowledge?

Given our purpose here is to understand increases in knowledge, it is also relevant to discuss what we mean by knowledge, as knowledge can be variously defined. Throughout our studies, we examined both *objective* and *subjective* knowledge. It should be noted that integration (Neuman, 1981) or structural knowledge (Eveland & Hively, 2009) is another important facet of knowledge, but one that is typically ignored in research on public deliberations. Objective knowledge is probably what most would think about when hearing the term knowledge—being able to correctly identify explicit pieces of information—that is, facts. Factual knowledge is used extensively in communication and political science literatures when trying to assess what people know about various topics. Within this definition of knowledge, it is assumed that information can be either correct or false. Consequently, the more correct pieces of information people are able to access, the more knowledgeable they are. In public engagement contexts, this would be assessed by measuring how much participants know about the topic at hand prior to participating in the engagement, and again following participation, or in our case at multiple time points throughout the engagement. Beginning with Study 3, we asked knowledge questions at four different time points, at the beginning of the semester (A1), just prior to receiving the reading material (pre-A2), just after reading (post-A2), and in the final assessment at the end of semester (A4). Participants' knowledge was not re-assessed immediately following the discussion activities (A3), so any additional learning from the discussion activities we would assume to see in the final assessment.

Objective knowledge is most often assessed by asking participants either true-false or multiple-choice questions. The literature on the extent to which individuals show large gains in knowledge during public engagements and via these measures is mixed at best. The majority does not show sharp increases in knowledge; however, there is some research that has found objective knowledge increases over time.

The form of knowledge that is more consistently found to be improved via participation in public engagements is subjective knowledge—how much individuals *feel* they know.

## 3.4   What Did They Learn?

Before we go into what improved participants' learning, it's worth noting that we had to continually refine our objective knowledge measures. For us, measuring subjective knowledge was relatively straightforward, and we used similar items throughout all five studies (see Chap. 2 for details). To create the objective knowledge measures, in Study 3, we created items based on information that was provided in a document participants were asked to read and then examined pre-post engagement statistics for each knowledge question in order to identify the questions that were most sensitive to detecting pre-post changes. While refining our knowledge questions, we learned a lot about what participants did *not* learn. For instance, one question we had that performed poorly was a "select all that apply" question about what were current areas of nanogenomic research and development. Of the five options, two options showed high levels of correct responses that did not vary. The other three options had very low correct responses initially and got worse over time. True-false questions also posed problems, both in that some showed very high levels of knowledge at pre-exposure and also some questions that participants continued to do poorly on even after completing the reading and the discussions. We determined that these questions were either too intuitive in the case of questions where individuals knew the answer before the public engagement activities or were too difficult to ascertain from the readings in cases where we did not see any improvements over time.

Looking at the means in Table 3.1, we see that our measures did detect knowledge changes. Using both objective and subjective measures, we consistently see increased knowledge scores over the course of the public engagement. The increases are most pronounced when examined just after the readings, but the knowledge gains persist when comparing the end of the semester to the initial or pre-reading measures. These increases in knowledge over time are statistically significant. The knowledge means were tested with repeated measures ANOVA, and in all cases, the later knowledge measures were significantly greater than the initial, pre-reading measures. In a few cases, there was a decrease in knowledge from just after the reading and the end of the semester, but that should not be surprising. Some loss of memory for facts over several weeks is to be expected. But even with that decrease, participants knew significantly more at the end of the semester than they knew at the beginning of the semester.

This suggests that there were some long-term objective effects of participating in this public engagement. That is, the public engagement as a whole did lead to citizens who were more knowledgeable about nanotechnology. This is also true for subjective knowledge. Our participants consistently reported that they felt they

**Table 3.1**  Knowledge means

| Objective knowledge | | | | |
|---|---|---|---|---|
| | Initial measure (A1) | Pre-reading (pre-A2) | Post-reading (post-A2) | End of the semester (A4) |
| Study 2 | | 0.46 (0.24) | | 0.69 (0.16)* |
| Study 3 | | | | |
| Multiple choice | 0.53(0.21) | 0.57(0.19) | | 0.71(0.19)* |
| T/F set A | | 0.76(0.20) | 0.79(0.17) | 0.73(0.21)* |
| T/F set B | | 0.56(0.17) | 0.62(0.19) | 0.61(0.19)* |
| Study 4 | | | | |
| Multiple choice | 0.49(0.23) | | 0.63(0.24) | 0.60(0.22)* |
| T/F set A | 0.65(0.33) | 0.70(0.13) | 0.79(0.12) | 0.81(0.30)* |
| T/F set B | | 0.66(0.16) | 0.74(0.17) | |
| Composite | 0.54(0.20) | 0.68(0.15) | 0.72(0.14) | 0.67(0.20)* |
| Study 5 | | | | |
| Multiple choice | 0.55(0.29) | | 0.79(0.28) | 0.69(0.27)* |
| T/F | 0.72(0.18) | | 0.82(0.19) | 0.83(0.14)* |
| Composite | 0.68(0.16) | | 0.81(0.17) | 0.81(0.13)* |
| Subjective knowledge | | | | |
| | Initial measure | Pre-reading | Post-reading | End of the semester |
| Study 3 | 1.67 (0.74) | 2.06(0.79) | 3.12(0.60)* | |
| Study 4 | 1.62(0.69) | 1.94(0.67) | 3.00(0.60) | 3.00(0.60)* |
| Study 5 | 1.64(0.72) | | | 3.06(0.56)* |

Note: Cell entries are means and standard errors in parentheses. The objective measure can be interpreted as a percentage of correct responses
*Indicates that the measure is statistically different from the measure obtained either in A1 or pre-A2, depending on when first measurement was, which reflects learning that would have occurred over the course of the semester. Post-reading scores (post-A2) would demonstrate learning that occurred during the reading, and A4 scores would capture learning from both the reading and the discussions

knew significantly more as the semester went on, with the most prominent effects observed after the reading. When we did have both post reading and end of the semester measures, we did not see significant increases between those time points. The group discussion occurred between the reading and the end of the semester, which suggests that reading the materials, rather than engaging in discussion with other participants, had the most profound effects on what they objectively knew and felt that they knew.

Knowing that the majority of the observed learning is occurring after the reading as opposed to after the discussion is critical from a theoretical perspective. When the learning happens in public engagements is frequently not explicitly addressed. Instead most scholars look at the effects of the engagement as a whole, regardless of what aspect most contirubuted to learning. Our findings here are consistent with others that show reading appears to have a stronger impact on learning (e.g.,

Muhlberger & Weber, 2006). Focusing on reading rather than discussion opens up a lot of possibilities when it comes to the ease of implementing engagement activities. It suggests that what is most important is the time to thoughtfully process new information; therefore remote deliberations, such as Becker and Slaton's teledemocracy (2000), are possibilities.

Next, we examine the effects of information presentation on how much our participants learned or felt they learned. How we presented the information had minimal to no effect on how much the participants learned. Critical thinking prompts were used in all four studies, and in all cases, the prompts did not result in higher objective or subjective knowledge. In Studies 3 and 4, the information organization manipulation was used, and pro-con versus topical organizational strategies also did not have an impact on objective or subjective knowledge. The interaction between information presentation and critical thinking prompts was also not predictive. These findings could be interpreted as optimistic or encouraging to practitioners. It appears to suggest that if the information presented meets some minimal deliberative expectations (e.g., we strove to use balanced information, although less balanced information in Study 5 did not appear to decrease learning), we do not need to spend much time or effort on how to best structure the arguments or further encourage deep processing. It appears that participants learn at similar rates regardless of the information presentation, and they do still learn during the public engagement event. Whether these strategies matter for other outcomes will be addressed in subsequent chapters, but from a knowledge perspective, simply providing relatively clear and balanced information is enough within this context. Our subjects and context are unique, but this provides an optimistic starting point for future studies.

We now turn to examining whether or not the extent to which participants were engaged with the materials influenced how much they learned, or felt that they learned, by the end of the public engagement event. Table 3.2 shows all significant relationships obtained between our engagement scales and the end of the study measures of knowledge. As you can see, the relationship between the various engagement scales and knowledge was much more apparent when examining subjective knowledge. There were fewer and less consistent relationships between objective knowledge and the various forms of engagement, though a couple measures proved more consistent. In general, it appears that individuals do learn, and how they engage with materials effects the extent to which they *felt* they have learned, but less so how much they objectively learn.

Looking first at objective knowledge, conscientious and open-minded engagement were positive predictors of learning, both significantly predicting learning in three of the nine different measures of knowledge. Disinterested and angry engagement were negatively related to learning. Disinterested engagement was a negative predictor of four of the nine measures of knowledge, whereas angry engagement was a significant negative predictor in three of the nine measures of knowledge. Active learning, social engagement, creative engagement, and close-minded engagement were never significantly associated with objective measures. On the whole, there is some evidence of engagement influencing learning, but not consistently.

Consistently across the three studies, conscientious engagement, active learning engagement, and creative engagement had a positive relationship with subjective

**Table 3.2**  Knowledge and engagement

**Objective knowledge**

|  | Conscientious | Openminded | Active learning | Social | Creative | Disinterested | Angry | Closeminded |
|---|---|---|---|---|---|---|---|---|
| Study 3 | | | | | | | | |
| Multiple choice | +.144 | | | | | -.175 | -.136 | |
| T/F set A | | | | | | | | |
| T/F set B | | | | | | | | |
| Study 4 | | | | | | | | |
| Multiple choice | | | | | | -.195 | | |
| T/F | | +.147 | | | | -.144 | | |
| Composite | | | | | | -.216 | | |
| Study 5 | | | | | | | | |
| Multiple choice | | | | | | | | |
| T/F | +.172 | +.161 | | | | | -.198 | |
| Composite | +.138 | +.161 | | | | | -.203 | |

**Subjective knowledge**

|  | Conscientious | Openminded | Active Learning | Social | Creative | Disinterested | Angry | Closeminded |
|---|---|---|---|---|---|---|---|---|
| Study 3 | +.282 | +.239 | +.272 | +.159 | +.294 | -.240 | -.143 | |
| Study 4 | +.239 | +.185 | +.202 | | +.156 | -.192 | | |
| Study 5 | +.202 | | +.312 | +.233 | +.201 | | | |

Note: Cell entries show the sign and value of any significant correlations between the type of knowledge and engagement measure. Engagement items in Study 2 were not identical to measures from Studies 3–5, and the full knowledge battery was not used until Study 3, which is why Studies 3–5 are shown here

knowledge. That is, those who indicated they were paying closer, more conscientious attention, actively and metacognitively engaging with the materials and were thinking creatively about the issues were more likely to indicate that they learned more over the course of the deliberative event. Social and open-minded engagement also occasionally had positive effects on reported subjective learning outcomes. The negative forms of engagement were not consistently associated with subjective knowledge. Disinterested and angry engagement were associated with less subjective knowledge in Study 3, but only disinterested engagement was related in Study 4. None of the negative engagement items were related to subjective knowledge in Study 5.

In summary, we can say that participants do learn and feel that they have learned, and how they report engaging with materials has a fairly consistent effect on how much they feel they have learned, but how we structured the materials does not have an effect on how they learn. This opens up the question of whether or not how we structure the materials influences how individuals engage. This may mean that rather than our experimental manipulations having a direct effect on knowledge, engagement could serve to mediate the effect of our experimental manipulations on knowledge.

## 3.5   What Mediates Knowledge?

Assessing the alternative information organizations used in Studies 3 and 4 indicates that the pro-con versus topical organization is not associated with any of the forms of engagement. However, there is a fairly consistent pattern for the critical thinking prompts in Studies 3 and 4. As noted in Chap. 2, beginning in Study 3 we used prompts that were less didactic and more gentle to "nudge" participants in the direction of critical thinking. Across Studies 3 and 4, we find consistently that conscientiousness, open-minded engagement, and active learning are positively associated with the critical thinking condition, while close-minded engagement is negatively associated with the critical thinking condition. This suggests that our critical thinking prompts resulted in more positive engagement with the pre-deliberation materials. Given the earlier presented findings indicating that positive engagement is associated with increases in knowledge, this suggests that the critical thinking condition has the potential to influence knowledge indirectly through increases in positive engagement and decreases in close-minded engagement. It should be noted, however, that the relationships between critical thinking and engagement observed in Studies 3 and 4 did not replicate with Study 5. There were no significant relationships between the critical thinking condition and the various forms of engagement in Study 5.

In order to further address the relationship between critical thinking and subjective knowledge, we employ a more stringent test of the relationships by using linear regression with controls and including all of the forms of engagement. The controls allow us to account for other variables that people might argue could account for the relationship between engagement and knowledge and include need for cognition, typical grades, gender, willingness to change their minds, and their prior familiarity with nanotechnology. Given that the relationship between engagement and subjective knowledge was the strongest, this is where we would expect to find significant relationships, which is indeed the case. These analyses indicate conscientious engagement positively predicts increases in subjective knowledge across Studies 3 and 4, even when including various controls and simultaneously examining the effects of the other forms of engagement. The regression also indicates that the critical thinking condition did not have a direct effect on subjective knowledge once the controls and varieties of engagements were considered. These relationships taken together provide evidence that critical thinking prompts might increase learning through influencing how individuals engage with the materials.

To examine if critical thinking had an indirect effect on knowledge, parallel mediation using Hayes (2013) PROCESS macro was utilized, and the conceptual figure is presented in Fig. 3.1. Parallel mediation via PROCESS allows all indirect relationships to be simultaneously tested in order to determine which variables are exerting the strongest influence on our outcome of interest. The results of the process model in study three indicate that two forms of engagement serve as significant mediators of the effect of being in the critical thinking condition on increases in subjective knowledge. That is, different types of engagement serve as the conduit of the influence of critical thinking prompts on subjective knowledge. Controlling for the same set of variables used in the regression analyses, and considering the potential mediating effects of all forms of engagement, there is a significant indirect effect of being in the critical thinking condition on perceived knowledge through increased levels of conscientious engagement and decreased levels of close-minded engagement. That is, being in the critical thinking condition increases conscientious engagement and decreases close-minded engagement, which are subsequently associated with perceived knowledge. In Study 4, conscientious engagement served as a significant mediator of the effect of being in the critical thinking condition and increasing perceived knowledge (utilizing the same analysis strategy as Study 3). Again, being in

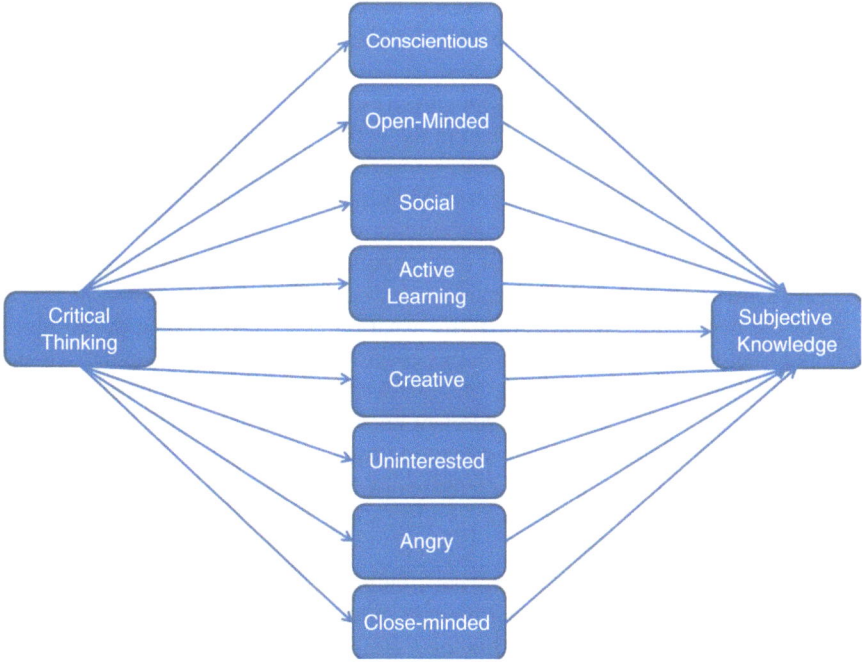

**Fig. 3.1** Conceptual figure for parallel mediation

the critical thinking condition increased conscientious engagement, which in turn was associated with higher levels of perceived knowledge at the end of the semester. Again, the indirect effects were not observed for objective knowledge.

## 3.6 Summary and Conclusion

Returning to our motivating question posed in Chap. 1 (what works, for what purposes, under what conditions, and why?), we were able to shed light on this question when it comes to addressing knowledge. In regard to addressing what works when it comes to knowledge, the combined influence of the public engagement activities appeared to have a positive effect. In particular, individuals both felt that they learned, in addition to showing objective increases in what they learned, after completing the readings that were provided. The structure of the readings was not found to be important in and of itself. Simply reading, at least within the specific public engagement we had set up, was enough to lead to increases in knowledge of what nanotechnology is and what it can do. The impact of discussion, regardless of the various scenarios we put participants in (see Chap. 2 for details), was minimal to nonexistent.

Finding the readings to be more important in comparison to the discussion is consistent with other research, despite being counter to what would be expected from deliberative theory. However, this issue is perhaps addressed by our results showing that cognitively engaging with the materials was associated with knowledge gains. Individuals who are more deeply processing the reading materials are essentially engaging in what deliberative scholars would desire, but doing it internally rather than with others. The benefits of discussion in deliberation are presumed to occur due to individuals thoughtfully considering issues and perspectives that are provided by other participants in the deliberation. By measuring how individuals were engaging with the content, we were able to determine that thoughtfully considering issues does in fact increase what individuals feel that they know, and this increase can happen in isolation rather than with others. Further study is needed to determine if this can occur out of a classroom setting or if these observed relationships are only found with students in classrooms.

The element of effectiveness (the "for what purposes" component of our motivating question) that was examined in this chapter was learning, conceptualized and operationalized two different ways. Making distinctions between what participants felt they learned and objective measures of knowledge was shown to be important. Subjective learning, but not objective learning, showed some of the more interesting relationships. These two different measures of learning are clearly different from each other as they frequently did not correlate significantly, although the sign of the coefficients were always positive.

The context element (the "under what conditions" component of our motivating question) was crucial to understand what we actually discovered about learning in public engagements about nanotechnology. How we structured the reading did not directly contribute to whether participant learned or felt they learned. This suggests that how you present information does not matter at all when it comes to objective learning. However, the critical thinking prompts did prove to be important once we considered how participants engaged with the readings. We find that the critical thinking prompts in Studies 3 and 4 encourage deeper processing and discourage negative forms of engagement, which subsequently lead to an increase in subjective knowledge. Including mediators that speak to *when* different elements may be more successful allowed us to paint a much more nuanced picture of what is occurring during public engagement activities.

From a broader perspective, having three studies that contained the full knowledge battery also allowed us to see what consistently worked or what consistently did not work. Looking at Table 3.2 paints a very clear picture of the importance of conscientious engagement—which is engaging in more thoughtful and deep processing—across all measures of knowledge and across all studies. Given the so-called replication crisis plaguing the social sciences, ensuring that we understand and appreciate the robustness of our effects is going to be important. Multiple studies were able to confirm the importance of conscientiousness and also show areas where effects may not be as robust as expected. In Study 5, the indirect effects of the critical thinking condition on subjective learning was not found, contrary to Studies 3 and 4. Study 5 had a different set of experimental manipulations in comparison to Studies 3 and 4, even though the critical thinking prompts remained the same. While we did not find direct or interactive effects of these various manipulations on learning or engagement, it is possible that these changes in the design of Study 5 changed the observed relationships. However, at this point, we have more findings than we have adequate theory to cover them.

# References

Allcott, H., & Gentzkow, M. (2017). Social media and fake news in the 2016 election. *Journal of Economic Perspectives, 31*, 211–236. https://doi.org/10.1257/jep.31.2.211

Becker, T. L., & Slaton, C. D. (2000). *The future of teledemocracy*. Westport, CT: Greenwood Publishing Group.

Bohn-Gettler, C. M., & Kendeou, P. (2014). The interplay of reader goals, working memory, and text structure during reading. *Contemporary Educational Psychology, 39*, 206–219. https://doi.org/10.1016/j.cedpsych.2014.05.003

Burkhalter, S., Gastil, J., & Kelshaw, T. (2002). A conceptual definition and theoretical model of public deliberation in small face—to—face groups. *Communication Theory, 12*, 398–422. https://doi.org/10.1111/j.1468-2885.2002.tb00276.x

Delli Carpini, M. X., Cook, F. L., & Jacobs, L. R. (2004). Public deliberation, discursive participation, and citizen engagement: A reivew of the empirical literature. *Annual Review of Political Science, 7*, 315–344. https://doi.org/10.1146/annurev.polisci.7.121003.091630

Delli Carpini, M. X., & Keeter, S. (1996). *What Americans know about politics and why it matters*. New Haven, CT: Yale University Press.

Druckman, J. N., & Bolsen, T. (2011). Framing, motivated reasoning, and opinions about emergent technologies. *Journal of Communication, 61*, 659–688. https://doi.org/10.1111/j.1460-2466.2011.01562.x

Dutwin, D. (2003). The character of deliberation: Equality, argument, and the formation of public opinion. *International Journal of Public Opinion Research, 15*, 239–264. https://doi.org/10.1093/ijpor/15.3.239

Eagly, A. H., & Chaiken, S. (1993). *The psychology of attitudes*. Belmont, CA: Wadsworth.

Eveland, W. P. J. (2001). The cognitive mediation model of learning from the news: Evidence from nonelection, off-year election, and presidential election contexts. *Communication Research, 28*, 571–601.

Eveland, W. P., & Hively, M. H. (2009). Political discussion frequency, network size, and "heterogeneity" of discussion as predictors of political knowledge and participation. *Journal of Communication, 59*, 205–224. https://doi.org/10.1111/j.1460-2466.2009.01412.x

Fishkin, J. S. (1997). *The voice of the people: Public opinion and democracy*. New Haven, CT: Yale University Press.

Garrett, R. K., Weeks, B. E., & Neo, R. L. (2016). Driving a wedge between evidence and beliefs: How online ideological news exposure promotes political misperceptions. *Journal of Computer-Mediated Communication, 21*, 331–348. https://doi.org/10.1111/jcc4.12164

Gastil, J., & Dillard, J. P. (1999). Increasing political sophistication through public deliberation. *Political Communication, 16*, 3–23. https://doi.org/10.1080/105846099198749

Goodin, R. E., & Niemeyer, S. J. (2003). When does deliberation begin? Internal reflection versus public discussion in deliberative democracy. *Political Studies, 51*, 627–649. https://doi.org/10.1111/j.0032-3217.2003.00450.x

Hart, P. S., & Nisbet, E. C. (2012). Boomerang effects in science communication: How motivated reasoning and identity cues amplify opinion polarization about climate mitigation policies. *Communication Research, 39*, 701–723. https://doi.org/10.1177/0093650211416646

Lukensmeyer, C. J., & Torres, L. H. (2006). *Public deliberation: A manager's guide to citizen engagement*. Washington, DC: IBM Center for the Business of Government.

Luskin, R. C., O'Flynn, I., Fishkin, J. S., & Russell, D. (2014). Deliberating across deep divides. *Political Studies, 62*, 116–135. https://doi.org/10.1111/j.1467-9248.2012.01005.x

Mansbridge, J. (1999). Everyday talk in the deliberative system. In S. Macedo (Ed.), *Deliberative politics: Essays on democracy and disagreement* (pp. 211–239). New York, NY: Oxford University Press.

Mill, J. S. (1860). *On Liberty* (2nd ed.). London, UK: John W.Parker & Son.

Muhlberger, P., & Weber, L. M. (2006). Lessons from the virtual agora project: The effects of agency, identity, information, and deliberation on political knowledge. *Journal of Public Deliberation, 2*, 1–41.

Neuman, W. R. (1981). Differentiation and integration: Two dimensions of political thinking. *American Journal of Sociology, 86*, 1236–1268. https://doi.org/10.1086/227384

Petty, R. E., & Cacioppo, J. T. (1986). The elaboration likelihood model of persuasion. In L. Berkowitz (Ed.), *Advances in Experimental Social Psychology* (pp. 123–205). New York, NY: Academic Press.

Powell, M., & Kleinman, D. L. (2008). Building citizen capacities for participation in nanotechnology decision-making: The democratic virtues of the consensus conference model. *Public Understanding of Science, 17*, 329–348. https://doi.org/10.1177/0963662506068000

Robinson, D. H., & Kiewra, K. A. (1995). Visual argument: Graphic organizers are superior to outlines in improving learning from text. *Journal of Educational Psychology, 87*, 455–467. https://doi.org/10.1037/0022-0663.87.3.455

PytlikZillig, L., Hutchens, M., Muhlberger, P., Wang, S., Harris, R., Neiman, J., & Tomkins, A. (2013). The varieties of individual engagement (vie) scales: Confirmatory factor analyses across two samples and contexts. *Journal of Public Deliberation, 9*(2). Retrieved from http://www.publicdeliberation.net/jpd/vol9/iss2/art8

Sanders, L. M. (1997). Against deliberation. *Political Theory, 25*, 347–376. https://doi.org/10.1177/0090591797025003002

Selin, C., Rawlings, K. C., de Ridder-Vignone, K., Sadowski, J., Altamirano Allende, C., Gano, G., … Guston, D. H. (2017). Experiments in engagement: Designing public engagement with science and technology for capacity building. *Public Understanding of Science, 26*(6), 634–649. https://doi.org/10.1177/0963662515620970

Shah, D. V., McLeod, D. M., Rojas, H., Cho, J., Wagner, M. W., & Friedland, L. A. (in press). Revising the communication mediation model for a new political communication ecology. *Human Communication Research, 43*, 491. https://doi.org/10.1111/hcre.12115

Sturgis, P., & Allum, N. (2004). Science in society: Re-evaluating the deficit model of public attitudes. *Public Understanding of Science, 13*, 55–74. https://doi.org/10.1177/0963662504042690

van der Linden, S. L., Leiserowitz, A. A., Feinberg, G. D., & Maibach, E. W. (2015). The scientific consensus on climate change as a gateway belief: Experimental evidence. *PLoS One, 10*(2), e0118489. https://doi.org/10.1371/journal.pone.0118489

# Chapter 4
# Attitude Change and Polarization

**Abstract** A key reason for conducting public engagements around science and innovation policies is to find out what the public thinks and feels about those policies and the innovations themselves. However, some scholars have suggested deliberation can create attitude polarization, which could be a barrier to effective group decision-making and social progress. Thus, it is important to know when, if, and why processes lead to polarization. In this chapter, we examine individuals' attitudes toward nanotechnology and describe whether and how they are impacted by the design of public engagement. We focus particularly on the degree to which individuals' attitudes change and perhaps become more extreme, as a function of deliberation. We find that for the most part, the average of participants' attitudes toward nanotechnological development shifted toward being slightly more cautious over the course of the semester during each study we conducted, although other significant patterns of attitude change were evident among individuals. The features of deliberation that most consistently influenced attitudes were critical thinking prompts and information formatting, such that encouraging critical thinking and presenting information in a way that presented multiple perspectives often led individuals to take on more cautious views toward nanotechnology. Other features commonly theorized as having important consequences for deliberation showed mostly no effects, and we found little evidence of attitude polarization, a phenomenon feared by many scholars who have remained skeptical of deliberation. However, the degree to which group dynamics during deliberative discussion (specifically, group homogeneity) influenced attitude change and polarization was moderated by the personality variable trait of openness. Those high in openness were the least likely to experience attitude extremitization (attitude change in the direction of becoming more extreme) in attitudinally heterogeneous groups but the most likely to experience attitude extremitization in attitudinally homogeneous groups.

**Keywords** Attitudes · Attitude change · Homogeneity · Heterogeneity · Benefits versus risks · Common ingroup identity model

**Electronic supplementary material**: The online version of this chapter (https://doi.org/10.1007/ 978-3-319-78160-0_4) contains supplementary material, which is available to authorized users.

## 4.1 Introduction

One of the main reasons researchers and public officials may want to conduct public engagements is to discover what the public actually wants when it comes to science and innovation policy. For many, scientific discovery and technological development may not necessarily seem like democratic ventures, but there are few scientific or technological advances that would have been possible without some degree of support from the public.

We need look no further than the case of genetically modified organisms (GMOs) in the USA and Europe to see how vital public support is to an emerging technology. GMOs are organisms whose genetic material has been altered in some way via genetic engineering, and they have been around since the 1970s (although some have argued that selective breeding, which has existed since 12,000 BC, is a form of genetic engineering; see Kingsbury, 2009). The applications of GMOs include medical research, the production of pharmaceutical drugs, development of biofuels, and plant and animal conservation, but the most controversial application of GMOs is in agriculture and food production. The scientific consensus is that GM foods are no more likely to cause harm to humans than other foods (e.g., Alessandro, Manzo, Veronesi, & Rosellini, 2013), and GM foods have provided nutrients for millions of people who otherwise would be severely malnourished (e.g., FAO). However, the public remains far more skeptical about GM foods than scientists in the USA and Europe (Funk & Rainie, 2015; Marris, Wynne, Simmons, & Weldon, 2001). This skepticism is surely driven in part by a lack of knowledge about what GMOs are, but it is also driven largely by individuals' values and moral learnings (Kam & Estes, 2016; Scott, Inbar, & Rozin, 2016). Further, some people's opposition to the use of GMOs in food seems driven not by concern over the health effects of GMOs but instead by ideological concerns such as perceived overreach by the government or the possibility of monopoly by large corporations invested in GMOs (Dewey, 2017). Ultimately, public opposition to GM foods has had substantial policy implications. Twenty-eight countries in the European Union (EU), as well as 36 other countries, require GM foods to be labeled as such, and 19 countries in the EU have "opted out" of growing GM crops. The US Congress recently passed a law mandating that information be made available regarding whether foods use GMOs (Charles, 2016). Laws mandating that GMO foods be labeled are not mere inconveniences for companies developing and using GMOs. These laws may have a significant impact on the consumption of GM foods in developed countries as well as on the distribution of GM foods to underdeveloped countries (The Economist, 2014), and proposals to ban the production of GM foods continue to spring up across the USA (Karlamangla, 2014).

Although the success of new technologies depends upon public acceptance and support (discussed further in Chap. 5), the lack of scientific knowledge and literacy among citizens (discussed in Chap. 3) makes it difficult for opinions to be developed and clearly expressed and for policymakers to decide how seriously to take public opinion in the first place. Even when citizens form opinions about science or technology, the issues often become politicized, leading to attitude extremity and polarization (McCright & Dunlap, 2011; Kahan et al., 2012; Lewandowsky, Gignac, & Oberauer, 2013). Members of the public often lack the familiarity with new technologies needed to grasp both its benefits and its risks, and views toward regulations are often politicized, hijacked by political rhetoric, and defined in extremist terms. Extreme views, which are far from unrepresented in the contemporary USA (e.g., McCarty, Poole, & Rosenthal, 2006), can yield polarization even over otherwise non-political issues. With polarization comes gridlock, which can stifle scientific and technological development as well as prevent policymakers from implementing effective regulations. For these reasons, scientists, investors, and policymakers are wise to be concerned with finding ways to measure public opinion toward science and technology, guiding development in a way that takes into account public opinion, and perhaps even developing engagement strategies that encourage citizens to adjust their attitudes based on new and accurate information.

As discussed in Chap. 1, scientists and policymakers have increasingly turned to public deliberations as a means of addressing a variety of concerns about democratic engagement, and among these concerns is the potential for polarization over controversial issues related to science and technology. The hope is that by getting citizens together, informing them, and having them hash out their differences, the public as a whole can come to a more enlightened, reasonable consensus and move forward accordingly. However, a substantial body of research in psychology, communications, and political science suggests we should question whether this is really what we should expect when citizens deliberate. It may be the case, for instance, that certain features of deliberation lead people to take sides, to become more extreme in their original views, or conversely even to acquiesce to a less informed, suboptimal opinion in response to conformity pressures.

In this chapter, we explore the effects of various features of deliberation on attitude change and polarization. The ways in which the features and context of a deliberation influence participants' views of the issue at hand should be a central concern of researchers and policymakers interested in scientific and technological development because these issues often have been shown to be easily politicized. Climate change, vaccinations, stem cell research, genetically modified organisms (GMOs), and even evolution are examples of science and technology issues that have been significantly impacted by public discourse, disagreement, and polarization. Properly designed public deliberation might represent an avenue for researchers and policymakers to avoid the pitfalls of a polarized public, but first we need to better understand how different features of deliberation influence people's attitudes.

The results of our analyses call into question some of the assumptions about the roles of various commonly used features of deliberation. The modal outcome of our experimental conditions is no effect on attitudes, although there are some cases in which we see indications of a pattern. The conditions that resulted in significant attitude change most often were those aimed at encouraging critical thinking in some way, but these effects were sporadic and did not occur in the majority of cases. These findings may be interpreted optimistically, because we did not find much evidence of adverse changes such as polarization or extremitization. However, we also did not find any evidence of conditions leading to "positive" outcomes. Counter to what many deliberative theorists would suggest, discussing nanotechnology in groups did not produce any significant changes in aggregate attitudes compared to simply deliberating about the issue on one's own, suggesting the positive impacts of discussion may be overstated.

## 4.2   The Effects of Deliberation: Unification or Polarization?

When is it that we should expect scientific and technological development to be welcomed with open arms versus shunned or even actively resisted with fear and skepticism? When should we expect deliberation to lead individuals to consensus versus polarization? What, if anything, should we expect to happen to people's attitudes when they are asked to deliberate about issues of science and technology? Over the last few decades, scholarship in psychology, communication, and political science has made some headway in shedding light on the answers to these questions. However, the conclusions of this scholarship have been somewhat mixed. Many of the large-scale deliberations conducted by scholars have shown substantial attitude changes via deliberation toward more well-informed opinions that resemble those of experts (e.g., Fishkin, Iyengar, & Luskin, 2005; Fishkin & Luskin, 1999; McLean et al. 2000), but it is often difficult to disentangle what exactly changed opinions, and quite different patterns of attitude change have been found in smaller-scale studies on the effects of group discussion. This suggests the relationship between deliberation and attitudes is more nuanced than simple analyses might imply. What we do know is that deliberations do not have a single, universal effect on people's attitudes. Context matters (Delli Carpini, Cook, & Jacobs, 2004), even if we do not yet fully understand why or how (Chap. 1). Below, we outline the existing theories and research that pertain to deliberation's "good" or "bad" effects on people's attitudes.

### 4.2.1   The Promises of Public Deliberation: Informed, Enlightened Consensus

The point of view that the possible benefits of deliberation outweigh the possible harms emanates predominantly from the theoretical arguments of deliberative theorists. Dewey (1927) argued that without the communication offered through public deliberation, apathy and self-serving biases would leave the public divided as citizens walled themselves off into disparate echo chambers. Although it is never argued that consensus will or should be the universal result of all deliberation, it is believed that exposure to new information and a diverse set of viewpoints through deliberation will or should lead to some degree of open-mindedness and engagement with alternative perspectives (see also, Chambers, 2003; Gutmann & Thompson, 1996;

Habermas, 1996; Fishkin & Luskin, 1999). At the least, according to some deliberative theorists, deliberation should lead people to come to terms with the idea that some level of disagreement is inevitable, and thus people will become more likely to tolerate opposing views (Cohen, 1998). In terms of individuals' attitudes, then, many deliberative theorists would suggest that deliberation gives people the tools to incorporate alternative opinions into their own.

Some research has shown that deliberation—or at least the exposure to information as part of the deliberation, as discussed in Chap. 3—increases factual knowledge and, thereby, presumably informed opinions. Across the world, Fishkin and Luskin have implemented "Deliberative Polls™," in which representative samples of the population are brought together to discuss public matters, question experts, and vote on critical issues. In the majority of cases, they have found evidence of increased knowledge and what seems to be well-informed consensus (Fishkin, Iyengar, & Luskin, 2005; McLean et al. 2000; see also, Price & Cappella, 2002). Other research suggests that participants may become more cooperative. Psychology research on small group discussions has demonstrated that face-to-face communication increases intragroup cooperation by allowing individuals to express their willingness to cooperate, gauge others' willingness to cooperate, and draw connections between their own interests and the group's interest (e.g., Bornstein, 1992; Bouas & Komorita, 1996; Sally, 1995; see Delli Carpini, Cook, & Jacobs, 2004) .

## 4.2.2 Deliberation's Downfalls: Motivated Reasoning and Polarization

The calls for skepticism regarding the effects of deliberation on attitudes are grounded primarily in social psychological theories. Skeptics of deliberation point to several psychological phenomena that suggest deliberation may do more harm than good. One psychological mechanism that runs counter to deliberative ideals is motivated reasoning, wherein individuals search for information that confirms their pre-existing beliefs in order to mitigate the cognitive dissonance (Festinger, 1957) that arises when information contradicts beliefs (Bodenhausen & Macrae, 1998; Schulz-Hardt, Frey, Luthgens, & Moscovici, 2000; Taber & Lodge, 2006). The concern this raises is that when individuals are exposed to alternative viewpoints through deliberation, they will double down on their pre-deliberation opinions, thus becoming more extreme in their views. This effect has been found in some studies that involved group or interpersonal discussion, albeit not structured deliberations (Mutz, 2006; Tetlock & Kim, 1987).

Other possible effects of deliberation that could be thought of as deleterious have also been considered by skeptics of deliberation. For example, some have cited research on group conformity pressures to suggest that although deliberation may cause individuals' attitudes within groups to move closer to one another, this consensus may be suboptimal if it is simply a reflection of the majority's pre-deliberation opinions and not influenced by new and relevant information (Isenberg, 1986; Myers et al., 1980). The "consensus" reached through discussion may even be disingenuous as individuals wish simply to avoid conflict and maintain a positive image in the group (Davis et al., 1989). Further, as opinions within groups conform to one another, this may lead to greater divergence between groups. Empirical evidence exists showing these effects can occur in some instances (Insko et al., 1993; Schkade, Sunstein, & Kahneman, 2000; see Muhlberger, Gonzalez, PytlikZillig, Hutchens, & Tomkins, 2017 for a summary of the different types of attitude change that may occur via deliberation).

## 4.3  What Works, for What Purposes, Under What Conditions, and Why?

In line with the framework set forth throughout this book, we proceed in this chapter by considering what works to impact attitudes in deliberation and why. We tracked changes in students' attitudes toward nanotechnology over the course of the semester and examined effects that our experimental manipulations had on these attitudes to understand what features had impacts on attitudes or attitude change and why.

### 4.3.1  For What Purposes?

We begin by considering the purposes for which the deliberative engagement is occurring, as this is a fairly subjective yet crucial decision that sets the tone for how a deliberation might be structured and how the data will be analyzed. In a broad sense, deliberative theorists have debated for decades whether and how attitudes "should" change as a function of deliberation, as described above.

Regardless of whether or not there is a desired direction for attitudes to shift via the deliberation, there are certain outcomes that are by and large seen as adverse. Most deliberations are not conducted with the goal of getting people to ignore alternative viewpoints and double down on their original opinions or getting people to come to a consensus around an extreme viewpoint that is uninformed or problematic in some way. As such, it is usually desirable not only to estimate the degree to which "desirable" attitudinal processes have occurred but also the degree to which "undesirable" attitudinal process have occurred. In this chapter,[1] we

---

[1] In a separate manuscript that uses the data from Study 5, we develop a statistical model for parsing out the distinct types of attitude change and polarization, which may be of use to those concerned about multiple possible outcomes of deliberation (Muhlberger, Gonzalez, PytlikZillig, Hutchens, & Tomkins, 2017). Although the model is easily estimable using OLS regression, the model remains somewhat involved, so we do not use it in this chapter.

present results from basic analyses that straightforwardly examine changes in attitudes as well as attitude extremitization as a function of time and experimental condition.

## 4.3.2    What Works, Under What Conditions, and Why?

The experimental manipulations used in our studies were chosen broadly to reflect commonly varied features of deliberations and to examine their impacts on our primary dependent variables of interest (in this chapter, attitudes). A variety of experimental manipulations have been used in other research to examine how attitudes change in response to deliberation or discussion occurring under different contexts. Perhaps most common are examinations of the effects of face-to-face discussion. As we have already described above, the outcomes of face-to-face discussion have been shown to vary, ranging from increased cooperation to polarization. In Studies 2, 3, and 4, we manipulated whether students discussed their views toward nanotechnology in groups or simply reflected upon their views alone. In doing so, we were able to experimentally test the degree to which face-to-face discussion influenced students' attitudes and, if so, how.

In our studies, we separated the time when students learned new information about nanotechnology from the time when they discussed nanotechnology in groups. As such, we were able to isolate, to some degree, one of the reasons *why* discussion has the effects that it does. Researchers disagree regarding whether the effects of group discussion on changes in people's attitudes are due to social influence (e.g., conformity pressures), to learning new information, or some mix of these two factors. If the effects of group discussion on attitudes are due to social influence, we should expect attitudes to change after group discussion. This would suggest those organizing a deliberation should make sure to either enhance or diminish social interaction accordingly. However, if the effects are due simply to new information, attitude change should be concentrated after students learn new information, but not after group discussion. The implication would be that group discussion may not be necessary for attitude change.

The composition of attitudes that exist within discussion groups has been found to matter substantially as well, when it comes to predicting changes in attitudes. For example, if a majority opinion exists, opinions will tend to move toward that pre-existing majority opinion (Schkade, Sunstein, & Kahneman, 2000). But other aspects of the discussion, like the degree to which group norms place value on original or innovative arguments (Moscovici, 1985) or whether the discussion is aimed at reaching a particular decision rather than simply having discussion for discussion's sake (Smith, Tindale, & Dugoni, 1996), can determine the relative impact of minority opinions (e.g., Bettencourt & Dorr, 1998; Maass & Clark, 1984; Moscovici & Mugny, 1983; see also, Delli Carpini, Cook, & Jacobs, 2004; Mendelberg, 2002). Further, Gaertner and Dovidio have shown under the framework of the *common ingroup identity* model that encouraging interaction between subgroups within a larger group can facilitate cooperation and reduce intergroup bias (Gaertner & Dovidio, 2014).

In Study 5, we varied a facet of attitude composition within groups that has been hypothesized to be central in some previous work (e.g., Mendelberg, 2002): attitudinal homogeneity. Specifically, all students participated in group discussion during Assignment 3 in Study 5, and we manipulated whether or not the groups were comprised of like-minded individuals in terms of attitudes toward nanotechnology—i.e., some groups were attitudinally homogeneous, and others were attitudinally heterogeneous. By manipulating the attitudinal homogeneity of discussion groups, we were able to shed light on the conditions under which group discussion might lead to one outcome (e.g., increased consensus) versus another (e.g., increased polarization).

We also used individual-level personality variables to investigate potential moderators of the effects we examined, to advance understanding of *why* or for whom attitudinal homogeneity within groups might matter. For example, it may be expected that individuals in attitudinally homogenous discussion groups are the most likely to become more extreme in their views after deliberation, due to homogenous discussion resulting in more closed-mindedness to other opinions that do not fit with the group's view. If that is the case, the effect might diminish among individuals who are high on openness to experience (i.e., individuals who are more amenable to the idea of changing their views based on exposure to alternative perspectives). This would suggest certain personality variables like openness are important to track during deliberations.

Aside from the different ways in which group discussion can occur, variations in how deliberation occurs at the individual level have also been shown to influence attitude change. For example, analytical thinking plays an important role because although individuals who tend to think analytically are more likely to deliberate and make valid arguments (Cacioppo et al., 1996), they are also more likely to resist alternative views (Petty et al., 1995). When individuals are made to feel accountable in some way for their decisions, they are more likely to evaluate information objectively and deliberate in an effortful manner (e.g., Tetlock & Kim, 1987). Finally, through the activation of particular emotional states, individuals can be encouraged to seek out new information and interaction with others (e.g., Marcus, Neuman, & MacKuen, 2000). These findings suggest that by changing the way individuals engage with and process information during a deliberation, it may be possible to change the way their attitudes are influenced. In all of our studies, we manipulated the degree to which students were encouraged to think critically throughout the study. These manipulations allowed us to investigate the role of deliberative, analytical thought in driving attitude change over the course of the semester.

## 4.4  Results

We describe our results regarding attitude change and polarization in two sections. In the first section below, we examine the trends in students' attitudes toward nanotechnology over the course of the semester across each of the four studies. In the second section, we examine the effects of the experimental manipulations used in each study. We examine three types of attitude change: attitude shifts (taking into account direction of change), absolute attitude change (attitude change regardless of direction), and attitude extremitization (attitude change in the direction of one's prior attitudes). We are particularly interested in absolute change and extremitization because these analyses give us some sense of the degree to which individuals are changing their minds and refining their opinions in general, as well as the degree to which attitude consensus versus polarization is occurring. The normative outcomes desired by most deliberative theorists entail some degree of attitude change— i.e., participants should be altering their attitudes based on new information learned during the deliberation; if their attitudes remain the same, then perhaps the expense of having a deliberation is wasted. However, if participants are simply becoming more extreme in their prior views, this would be counter to the ideals of most deliberative theorists—hence the importance of measuring extremitization.

We focus here on our broad measures of students' attitudes toward nanotechnology, as these measures are the most consistent across studies. Specifically, all studies contained an item measuring the degree to which students believed the benefits of nanotechnological development outweigh the risks or vice versa (using a Likert scale ranging from 1 to 5 or 1 to 7), as well as an item measuring how much regulation or deregulation students believed there should be regarding nanotechnological development (measured on a sliding scale from 0 to 100). Various other, more specific attitude measures are available within each study (see Chap. 2 and Supplementary Materials) but are not analyzed here.

We keep our analyses simple. To examine attitude shifts over the course of the semester, we use paired sample $t$-tests. We use one-way ANOVAs to gauge whether mean differences between experimental conditions are significant. We transformed student responses to the two attitude items into six dependent variables: Mean attitudes were simply the average score across individuals on each attitude item. Mean absolute change for each item was calculated as the mean absolute value of attitude change from the time of the manipulation to a given measure administered later (i.e., the average amount of attitude change regardless of direction). Finally, mean levels of extremitization, or movement in the direction of one's prior attitudes, were calculated the same way as mean absolute change except that movement in the same direction as one's prior attitude score (i.e., movement away from the midpoint of the scale) was positive and movement in the opposite direction of one's prior attitude score (i.e., movement toward or even past the midpoint of the scale) was negative (for individuals whose prior attitudes were exactly at the midpoint of the scale, movement in either direction was coded as positive). Overall, significant interactions between conditions were rare in our data and fairly weak when they did exist. We therefore present main effects.

### 4.4.1 Attitude Change over Time

Figure 4.1 illustrates mean attitude scores over the course of the semester for each study for both of our primary attitude items. Across all studies, students started with fairly optimistic views toward the benefits versus risks of nanotechnology but also

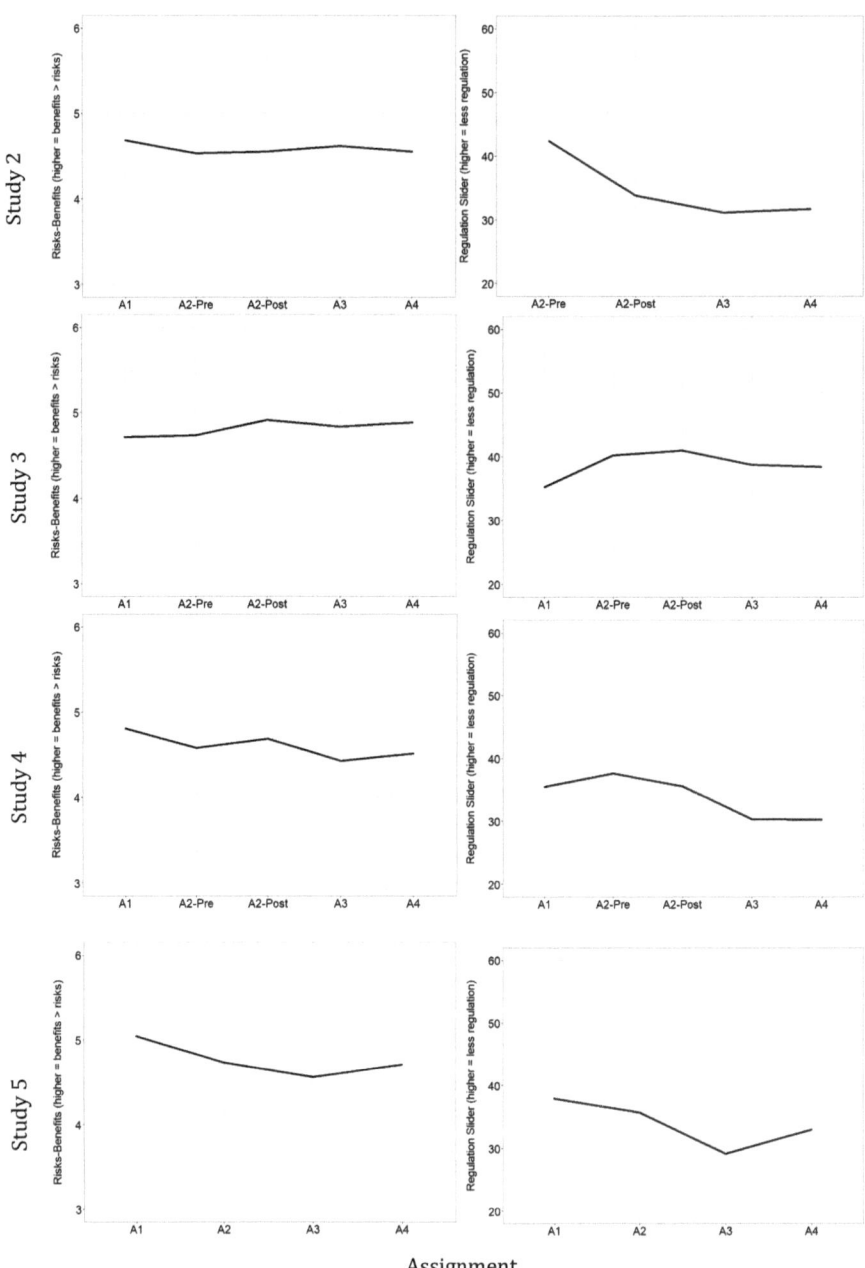

**Fig. 4.1** Attitude shifts over the semester. Greater values on the y-axis for "Risks-Benefits" indicate greater valuation of the benefits over the risks of nanotechnological development; greater values on the y-axis for "Regulation Slider" indicate preferences for less regulation and more development of nanotechnology; for Assignment 1 in Study 2, the response options ranged only from 1 to 5 and so were rescaled to range from 1 to 7 such that 2, 3, 4, and 5 were recoded to 3, 4, 5, and 7, respectively; the Regulation Slider item was not asked during Assignment 1 in Study 2

generally favored regulation of nanotechnology rather than development (higher scores on the deregulation items indicate support for fewer regulations). Over the semester, significant changes in aggregate attitudes occurred but were modest. In all studies except Study 3, students generally became more cautious toward nanotechnology over the course of the semester, placing more weight on the risks (vs. benefits) of nanotechnological development and becoming more supportive of regulation (vs. deregulation).

A closer look at the changes over time indicated that, although the overall pattern was toward caution, in most cases, students first moved toward regulation and then rebounded toward deregulation by exhibiting a statistically significant shift between the last two attitude measures.[2] The shifts during the first few assignments and then the slight increase in a less cautious direction between the last two assignments could be interpreted as reflecting deliberative quality: as students learn more about nanotechnology, the allure of new technology may be somewhat eclipsed by new knowledge that there are risks involved. Then, as students have more time to think about the issues, they rebound a bit—taking on more moderate stances. However, the overall shift was still significant and toward feeling more cautious toward nanotechnology. Study 3 is the exception because students steadily became *less* cautious toward nanotechnology across all assignments. Nonetheless, across studies, aggregate attitude changes were by no means drastic.

Despite these trends, looking at aggregate patterns can be misleading. It could be that substantial attitude changes occurred in individuals, but the changes canceled out on average across persons. As such, we look to Fig. 4.2 for an illustration of how the mean absolute change in attitudes varied across each semester. Even when examining absolute change, there was no case in which we saw evidence of drastic attitude change. In all cases, mean levels of absolute change were low. Differences between time points within semesters, though, tended to be statistically significant, suggesting some time points exhibited significantly more change than others. In general (except for absolute change regarding the deregulation item in Study 2 and the benefit item in Study 3), most of the attitude change that occurred tended to take place between the first two assignments—that is, *prior to* being given any information about nanotechnology. This suggests that counter to what many deliberative theorists would consider "optimal," the largest one-time attitude changes occurred between the time people were initially exposed to the topic (when they were asked questions about nanotechnology and told they would be informed about and discuss the topic later) and actual exposure (possibly reflecting self-seeking of information) rather than during the main deliberative activities. According to the results in Studies 2 through 4 (except for the cases mentioned above), most participants changed their

---

[2] This pattern held for all cases except in Study 3 and in Study 2 with regard to the risks versus benefits of nanotechnology.

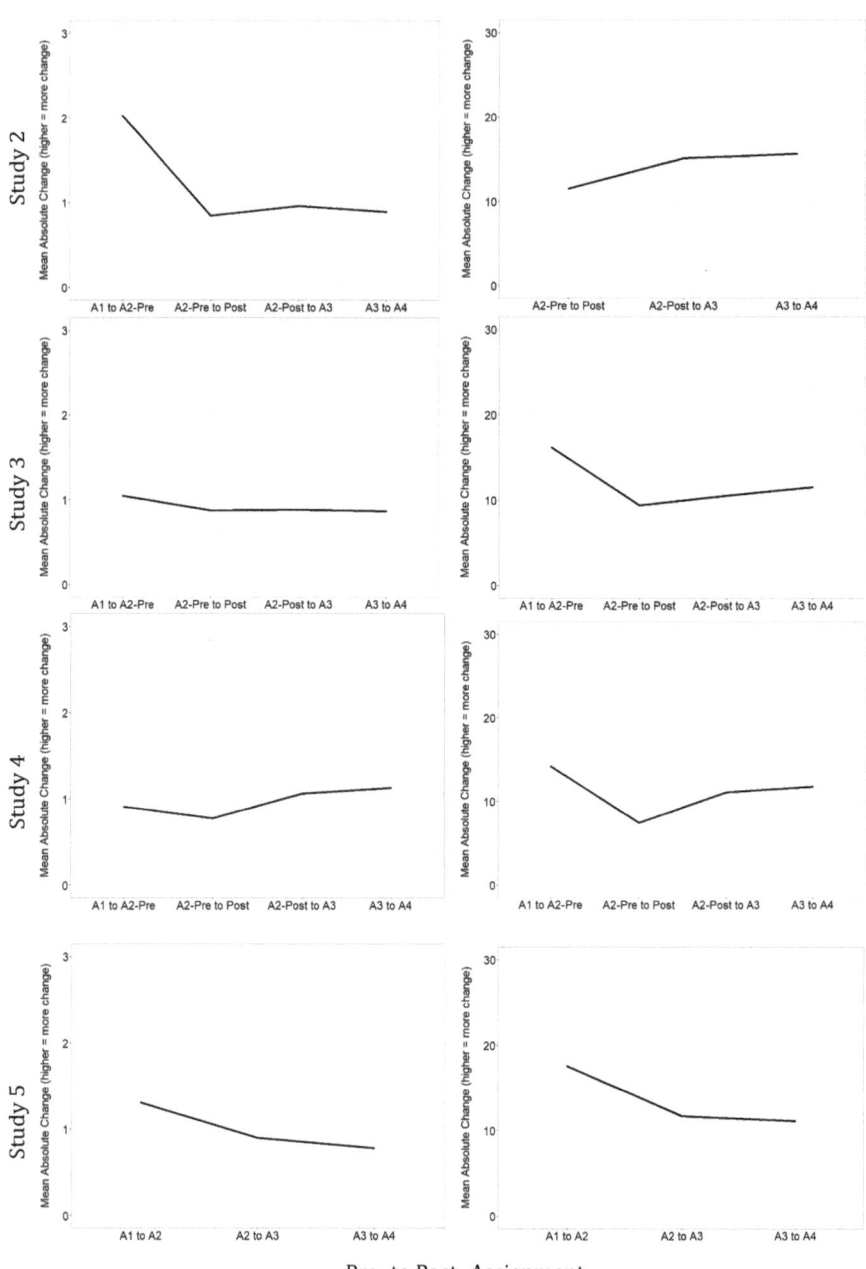

**Fig. 4.2** Mean absolute change in attitudes over the semester. Greater values on the y-axis indicate greater levels of attitude change from the prior Assignment (regardless of direction); for Assignment 1 in Study 2, the response options ranged only from 1 to 5 and so were rescaled to range from 1 to 7 such that 2, 3, 4, and 5 were recoded to 3, 4, 5, and 7, respectively; the Regulation Slider item was not asked during Assignment 1 in Study 2

minds even before they were given detailed information about the topic or had the chance to discuss the topic with others in the context of our study.

Although the absolute change in attitudes during the deliberative activities was small, we would caution against too much pessimism for a number of reasons. First, although not overwhelming, the mean level of absolute change in attitudes across all time points and all studies is above zero, suggesting people are not entirely static in their opinions during the deliberative activities. Also, it is important to note here that low levels of attitude change are not necessarily counter to deliberative ideals, as some have argued that what is important is the development of respect for alternative opinions rather than changing one's own opinion (Cohen 1998). Arguably more important is the degree to which extremitization—the most deleterious potential outcome of deliberation—did or did not occur in our data. We turn to that next.

Figure 4.3 shows mean levels of attitude extremitization across our studies. These values were computed by averaging the extent to which people changed their attitudes toward a more extreme view relative to their last reported attitude (resulting in positive values) or toward a more moderate or opposite view relative to their last reported attitude (resulting in negative values). Zero reflects no change in attitude. Our data does not show substantial levels of attitude extremitization at any time point during any of our studies. Average levels of extremitization tended to hover around zero, suggesting that the extent to which individuals became moderate in their opinions was at about the same level as others became more extreme. When significant changes did occur in extremitization, it was such that students became more likely to moderate their opinions, not that they became more extreme. That is, there were several instances in which extremitization scores went from below zero at the beginning of the semester (indicating movement toward more moderate views) to approximately zero by the end of the semester (indicating no further movement).

Of course, it could still be the case that differences in attitudes, absolute change in attitudes, or extremitization varied by different conditions. As such, we turn to our analyses regarding the effects of our experimental conditions to obtain a clearer picture of how people's attitudes changed in our data.

### 4.4.2  Encouraging Critical Thinking

A key factor in any public deliberation is the degree to which participants engage in critical thinking. On the one hand, effortful thought, scrutiny, and consideration of alternative viewpoints are often believed to be essential to successful deliberation, in part because critical thinking should (according to pro-deliberation theorists) lead people to think more objectively. In some cases, this might mean individuals become aware of the weaknesses of their own views and strengths of others' views and

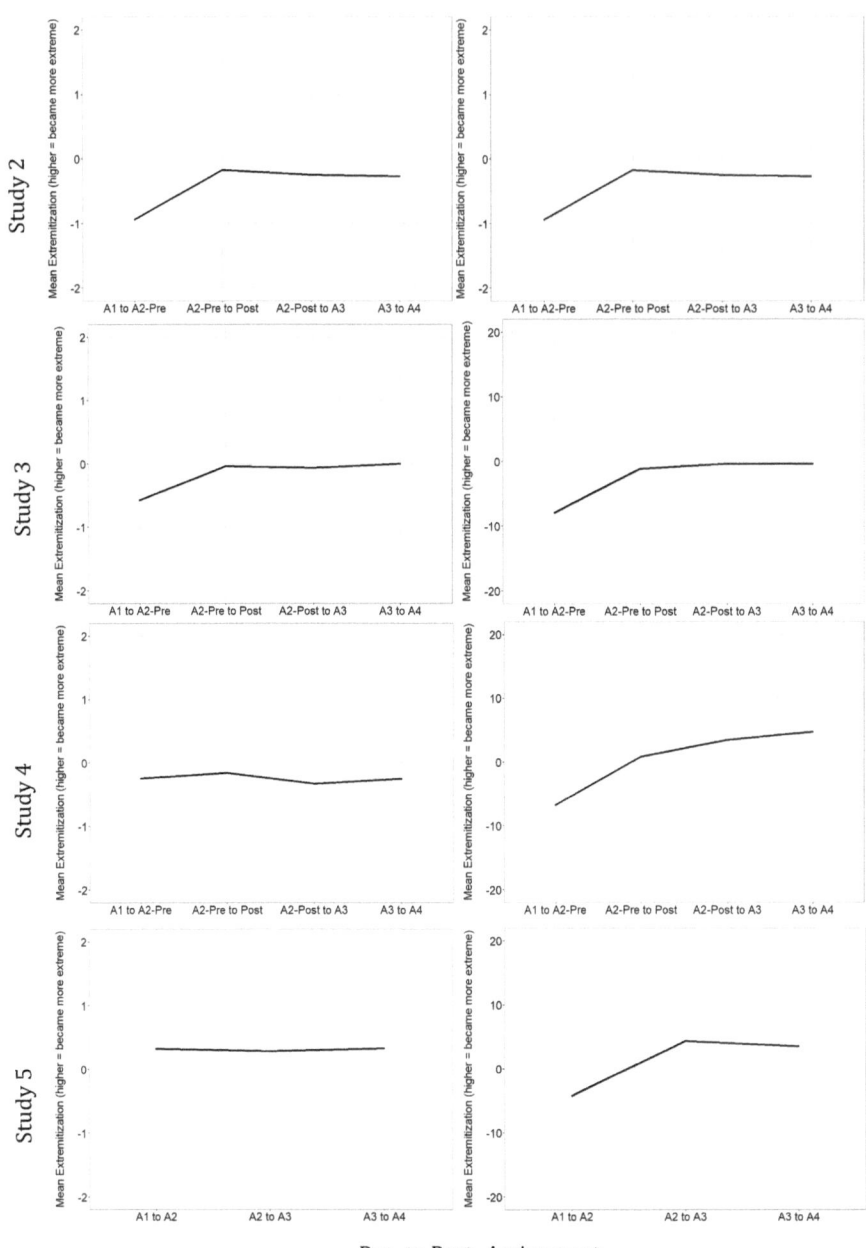

Pre- to Post- Assignment

**Fig. 4.3** Mean attitude extremitization over the semester. Greater values on the y-axis indicate greater levels of attitude change from the prior Assignment in the direction of being more extreme in the same direction as their attitudes in the prior Assignment (positive values indicate movement to a more extreme position and negative values indicate movement in the opposite direction; in cases where one's previous attitudes were at the exact midpoint of the scale, movement in either direction is considered extremitization and thus receives positive values); for Assignment 1 in Study 2, the response options ranged only from 1 to 5 and so were rescaled to range from 1 to 7 such that 2, 3, 4, and 5 were recoded to 3, 4, 5, and 7, respectively; the Regulation Slider item was not asked during Assignment 1 in Study 2

subsequently moderate their opinions. In other cases, research as well as sheer logic may come down concretely on one side of an issue, and so critical thinking may lead people to a specific point of view, which may even lay at the "extreme" end of the attitudinal spectrum. This may be desirable if the goal of the deliberation is more informed opinions rather than moderate opinions.

On the other hand, some research suggests critical thinking might lead to increased polarization rather than consensus. The argument here is that instead of leading to cool, open-minded consideration of alternative viewpoints, critical thinking may involve rationalization of one's previously held views and thus higher levels of attitude extremity (e.g., Kahan et al., 2012).

Table 4.1 contains the results of our analyses examining the effects of critical thinking prompts on attitude change and polarization regarding both dependent variables. Results are shown across all four studies and across all assignments following the manipulation within each study. That is, for each study the table shows the mean attitude score, mean absolute change in attitudes, and mean attitude extremitization during each assignment subsequent to the first administration of the critical thinking or alternative conditions (the first administration of the critical thinking manipulation is Assignment 2 in all studies). Note that in Study 2, there was an additional experimental condition beyond just the critical thinking and feedback (control) conditions called the information organization condition (mentioned in Chap. 2).

We start by describing the results regarding absolute change in attitudes and attitude extremitization because those are the most related to our expectations from the existing literature. In terms of differences in absolute change in attitudes, the modal outcome was no significant differences between conditions (this was the case in 66 out of 72 or 92% of comparisons). However, when significant differences did exist, they were mostly consistent. In Study 4 during Assignment 2-Post and Assignment 3 (for both the benefit item and the deregulation item) and in Study 5 during Assignment 4 (for just the deregulation item), being in the critical thinking condition was associated with *less* absolute change in attitudes. Thus, the majority of cases exhibited no significant differences between the critical thinking and feedback conditions, but when differences did arise, they suggested people were less likely to change their minds if exposed to the critical thinking prompts.

With regard to extremitization, the modal outcome was again no significant differences between conditions (just as with absolute change, this was the case in 66 out of 72 or 92% of comparisons). The majority of significant differences were evident in Study 3. Across all assignments in Study 3, being in the critical thinking condition was associated with attitude moderation (i.e., moving toward or past the midpoint on the scale) in terms of weighing the risks against the benefits of nano-technological development, whereas being in the feedback (control) condition was associated with extremitization. There were a couple of marginal ($p < 0.10$) differences in the opposite direction in Study 4 (where those in the feedback condition moderated their opinions but those in the critical thinking condition did not), but the opposite effects were never as large as those found in Study 3, and critical thinkers never were found to become more extreme.

**Table 4.1** Effects of critical thinking on attitude change and extremitization

| Study | | Condition | Risks-benefits | | | Regulation slider | | |
|---|---|---|---|---|---|---|---|---|
| | | | Mean | Mean absolute change | Mean extremitization | Mean | Mean absolute change | Mean Extremitization |
| Study 2 | A2-post | Feedback | 4.89 | 0.65 [a^] | -0.05 | 31.45 | 12.48 | 4.52 [a^] |
| | | Critical thinking | 4.47 | 0.86 | -0.08 | 33.59 | 9.78 | 5.34 |
| | | Information organization | 4.33 | 1.02 [a^] | -0.36 | 36.13 | 12.17 | -1.27 [a^] |
| | A3 | Feedback | 5.04 [a^a] | 0.75 | -0.02 | 29.05 | 15.27 | 3.71 |
| | | Critical thinking | 4.36 [a^a] | 1.02 | -0.22 | 31.36 | 14.69 | 5.10 |
| | | Information organization | 4.48 | 1.10 | -0.48 | 32.47 | 15.31 | -1.95 |
| | A4 | Feedback | 4.96 [a] | 0.69 | -0.09 | 29.24 | 17.06 | 3.24 |
| | | Critical thinking | 4.21 [a] | 0.96 | -0.31 | 31.21 | 14.40 | 5.25 |
| | | Information organization | 4.48 | 1.02 | -0.39 | 34.16 | 15.44 | -2.26 |
| | A5 | Critical thinking | 4.19 | 1.01 | -0.22 | 33.93 | 13.69 | 0.51 |
| | | Information organization | 4.48 | 1.02 | -0.34 | 34.47 | 14.98 | -2.68 |
| Study 3 | A2-post | Feedback | 5.06 [^] | 0.90 | 0.18* | 58.28 | 9.48 | -1.45 |
| | | Critical thinking | 4.77 [^] | 0.84 | -0.24* | 59.65 | 9.30 | -0.83 |
| | A3 | Feedback | 4.93 | 0.89 | 0.16* | 60.38 | 10.97 | 0.32 |
| | | Critical thinking | 4.75 | 0.88 | -0.29* | 62.10 | 10.09 | -1.00 |
| | A4 | Feedback | 5.08 [^] | 0.87 | 0.27* | 61.19 | 12.32 | 0.02 |
| | | Critical thinking | 4.71 [^] | 0.87 | -0.24* | 61.89 | 10.77 | -0.70 |

| Study | | Condition | Risks-benefits | | | Regulation slider | | |
|---|---|---|---|---|---|---|---|---|
| | | | Mean | Mean absolute change | Mean extremitization | Mean | Mean absolute change | Mean Extremitization |
| Study 4 | A2-post | Feedback | 4.71 | 0.93* | −0.32^ | 63.13 | 9.36* | 0.68 |
| | | Critical thinking | 4.67 | 0.63* | −0.01^ | 65.47 | 5.70* | 0.98 |
| | A3 | Feedback | 4.36 | 1.29* | −0.52^ | 69.05 | 13.95* | 3.07 |
| | | Critical thinking | 4.51 | 0.86* | −0.16^ | 70.48 | 8.59* | 3.86 |
| | A4 | Feedback | 4.60 | 1.26 | −0.34 | 68.84 | 14.30 | 5.34 |
| | | Critical thinking | 4.52 | 1.01 | −0.16 | 70.80 | 9.48 | 4.13 |
| Study 5 | A2-post | Feedback | 4.82 | 1.24 | −0.67 | 35.94 | 16.21 | −2.94 |
| | | Critical thinking | 4.65 | 1.38 | −0.72 | 35.54 | 18.83 | −5.56 |
| | A3 | Feedback | 4.71 | 0.96 | −0.45 | 28.60 | 11.85 | 1.66 |
| | | Critical thinking | 4.39 | 0.83 | −0.43 | 30.08 | 11.53 | −1.30 |
| | A4 | Feedback | 4.88 | 0.88 | −0.42 | 34.42 | 13.16* | −3.01 |
| | | Critical thinking | 4.59 | 0.69 | −0.17 | 32.18 | 9.24* | −0.85 |

Higher values for the "Risks-Benefits" column indicate greater valuation of the benefits over the risks of nanotechnological development. Higher values for the "Regulation Slider" column indicate preferences for less regulation and more development of nanotechnology; "Mean Absolute Change" indicates the absolute value of attitude change from A2-Pre to that assignment. "Mean Extremitization" indicates change in the direction of one's A2-Pre attitude score, with positive values indicating movement to a more extreme position and negative values indicating movement in the opposite direction (a score of 4 out of 7 indicates the midpoint for the Risks-Benefits item, and a score of 50 indicates the midpoint for the Regulation Slider item); in cases where one's A2-Pre attitudes were at the exact midpoint of the scale, movement in either direction is considered extremitization and thus receives positive values; mean differences were calculated using between-groups $F$-tests and pairwise comparisons with the Tukey HSD standard for statistical significance when there were more than two groups; * next to values indicates differences that are significant at the $p < 0.05$ level; ^ indicates differences that are significant at the $p < 0.10$ level; when there are more than two conditions, superscripts are used such that conditions with the same letter are significantly different from one another

When considering directional attitude change, for the most part, the critical thinking manipulations did not have an effect on attitude shifts in a particular direction. However, there are several instances of significant effects, and the direction of the effect is consistent in all cases but one. During Assignment 4 in Study 3, being in the critical thinking condition led to more negative attitudes toward nanotechnology (believing the risks outweigh benefits) than being in the feedback condition. This same relationship was evident in Assignments 2 and 4 of Study 3. However, average attitudes toward regulation of nanotechnology were never affected by the critical thinking manipulations.

Overall, then, any effects of critical thinking were somewhat sporadic, but when there were differences, participants sometimes became more negative yet while showing less attitude movement in the critical thinking conditions. Results generally suggest the primary effect of our critical thinking prompts on attitude change and variation is potentially moderation of attitudes but, more often, no effect. Despite not suggesting a single, widespread effect of critical thinking, our findings are notable. Our findings suggest that critical thinking is not universally causing people to refine (change) their opinions as deliberative theory purports, but it is also not causing polarization or extremitization.

## 4.4.3   Information Format

The formatting of information read by participants in a deliberation may seem like a trivial matter when it comes to their attitudes toward the topics at hand, but substantial variation exists across public deliberations in how information is presented, if at all. Researchers and policymakers—especially those focused on science and technology issues—often seek not only to measure public opinion via deliberations but also to inform and potentially guide it. As such, it is particularly common during deliberations over science and technology issues for information to be provided to participants that gives them a basic understanding of the topic. A reasonable concern for those who organize deliberations is how the *ways* in which they present information to participants might shift their opinions.

Throughout our studies, Assignment 2 served as a time for students to read background information about nanotechnology and nanogenomics, and in Studies 3 through 5, we manipulated the information as described in Chap. 2. The manipulations used in Studies 3 and 4 had to do with whether or not the risks and benefits of nanotechnological development were shown as alternative perspectives (pro-con condition) or simply in paragraph form without any clear division into opposing perspectives (topical condition). These differences in formatting have clear practical relevance, as the pro-con formatting was based off of the formatting used by the *National Issues Forum,* an organization whose specific aims include encouraging a shared understanding of issues across diverse views. It is often believed by many deliberation practitioners that by directly exposing people to opposing viewpoints, they will see the potential weaknesses in their own views and strengths in others' views, and this will improve the quality of deliberation (Gutmann & Thompson, 1996, 2009; Habermas, 1989, 1996; Fishkin, 1991). However, it may also be possible that by explicitly presenting issues as split into groups of opposing views, the background information may be politicizing the issues and making it easier for participants to become polarized. Group distinctions may become more salient, which may diminish the potential for compromise (Bettencourt & Dorr, 1998).

As discussed in Chap. 2, Study 5 used the NIF format for all participants and implemented a weak versus strong background information manipulation, which varied the degree to which sources and evidence were provided to back up arguments, the use of opinion-based claims, and the extent to which the information was balanced. This manipulation allowed us to test a fairly straightforward set of competing hypotheses. On the one hand, it could be the case that overly positive information, despite being poorly supported and stated, led to more positive views toward nanotechnology. On the other hand, the overly positive information could lead to a backlash effect because of it being weakly supported, with participants becoming more cautious.

The results regarding information format in Studies 2 through 4 were sporadic. Some differences existed, but, for the most part, there was not a consistent effect of the information format. The most consistent results regarding information format occurred with regard to attitude shifts during Study 5, when the information varied in terms of "strength."

In Study 5, the strong information condition consistently led to greater concern about the risks of nanotechnology. A reasonable interpretation of this pattern of results might be that conditions in which people were exposed to more balanced and well-supported information led to greater concern regarding the risks of nanotechnology. This suggests the deliberative ideals of balanced, unbiased information may not necessarily yield polarization but may nonetheless lead individuals to view new technology more cautiously. However, none of these effects carried over to mean attitudes regarding deregulation. There were only small and inconsistent differences associated with mean absolute attitude change.

There are various other information formats that might have different effects on attitudes, and results may differ further depending on the topic. Here, the topic (nanotechnology) was fairly novel for most participants (see Chap. 3), and so it should be expected that the background information would have a substantial impact on how participants formed their opinions toward the matter. Several significant shifts were evident, and there were some differences by formatting condition that were somewhat telling. If anything, framing the issue in terms of opposing perspectives rather than using a more topical approach to laying out benefits and risks had a conservatizing effect on attitudes in the aggregate. However, these effects were small. This leaves room for examining possible mediators or moderators of these effects or for studies looking at why attitudes toward a novel issue like nanotechnology would not be substantially impacted by learning about the topic.

## *4.4.4  The Effects of Group Discussion*

Group discussion is a central feature of deliberations and is particularly relevant to expectations regarding attitude change and polarization. Indeed, various scholars in psychology, communications, and political science have studied the effects of group discussion on attitude change, some even focusing specifically on the implications for public deliberation. All in all, the results of empirical work on the topic are mixed and suggest a range of possible outcomes of group discussion during deliberation as described earlier.

In our first few studies, we sought simply to examine if attitudes differed as a function of whether or not students discussed the issues with a group or not. As such, in Studies 2 through 4, we randomly assigned some students to discuss ethical scenarios related to nanotechnological development in groups and others to consider the ethical scenarios alone, on their own. In Study 3, some students were also placed in one of the two conditions using an online wiki forum, which we used simply as pilot data due to the lack of random assignment (see Chap. 2).

Surprisingly (given the extant literature on the subject), we found no significant effects on attitudes of being in a group versus being alone when considering the ethical scenarios except for a few marginal and contradictory differences. This could suggest that the primary attitude changes that occur during consideration of the ethical aspects of scientific and technological development are due mainly to thinking about the issues prior to discussion, rather than during discussion with others. This would cast some doubt over claims about the power of social influence over people's attitudes, at least when it comes to deliberation about science and technology. On the other hand, discussion with others may play a role in motivating people to read and consider new information in more ordinary contexts in which people cannot be told to sit and think about an issue.

## *4.4.5  The Features of Group Discussion: Homogeneity
##          and Facilitator Activity*

In Study 5, we wanted to delve into the features of group discussion that might affect attitudes and polarization. Although there were no significant differences between the alone and group conditions in Studies 2 through 4, group discussion is a central part of many public deliberations, and so we wanted to further explore if particular features of a group discussion affect participants' attitudes. Therefore, in Study 5, all students discussed the ethical scenarios of Assignment 3 in groups. We manipulated two features of the discussion: the attitudinal homogeneity of the group and the activities of the discussion facilitators who were instructed to lead the group in an active or passive manner as noted in Chap. 2.

Manipulating the homogeneity of the group was directly inspired by the existing psychology literature on small group discussions. A central aspect of the scholarly disagreement over whether or not deliberation will lead to consensus rather than polarization has to do with the psychological consequences of encountering viewpoints that differ from one's own. To put it simply, those optimistic about the effects of deliberation suggest that being exposed to alternative viewpoints will lead people to develop an appreciation for other opinions, thus yielding lower levels of variability in attitudes, whereas those less optimistic about the effects of deliberation suggest people will become resistant and double down on their original opinions, thus yielding increased polarization and variability. Homogeneity, alternatively, may lead group members to reinforce one another's pre-existing opinions, or it may lead group members to realize the "one-sidedness" of their group's opinions and search for alternatives.

Surprisingly, across assignments, there were no cases in which attitudes, absolute change, or extremitization differed significantly across discussion conditions. Prior research suggests that, at the least, attitudes should move around more in heterogeneous groups, either because individuals are attending to alternative viewpoints and coming to more reasoned opinions or because individuals are doubling down on their original opinions (thus becoming more extreme in the direction of their original opinions). We found no evidence of extremitization *or* increased attitude movement in general when individuals were in heterogeneous versus homogeneous groups. Optimistically speaking, this means that we find no evidence of what has been feared by many skeptics—that is, polarization via motivated reasoning and resistance to alternative views. Yet this also means that deliberative theorists' hopes that exposure to alternative perspectives will lead people to acknowledge others' opinions and change their minds also are unrealized in our data.

The manipulations regarding the role of discussion facilitators were driven more by practical concerns. There is substantial variation in whether or not public deliberations utilize discussion facilitators, and among those that use facilitators, there is substantial variation in how those facilitators are instructed to guide discussion (if they are instructed at all). As such, we sought to shed light on the ramifications of an active facilitator relative to one who steps aside and lets participants guide the discussion.

In our data, we found no evidence of attitude differences between the passive and active facilitator conditions. Taking into account the null results regarding group homogeneity as well as the manipulations of whether or not students discussed the topics in a group at all, our findings regarding group discussion seem quite straightforward. We seem to be left with astonishingly little support for the hypotheses derived from existing literature.

## 4.4.6 A Potential Moderator of Homogeneity

Before we conclude that the dynamics of group discussion have no meaningful effects on attitudes, we briefly examine whether some aspect of personality might play a significant moderating role. In line with the overarching framework of this book, we would like to emphasize that although we observed only minor evidence of attitude change and extremitization in our data in the aggregate, and although most of the effects of the experimental conditions were either weak, inconsistent, or insignificant, it is possible that we may have missed something by averaging over participants. As such, we look at how personality—specifically, openness—might have played a role in driving our findings. We focus particularly on the potential moderating role of openness on the effects of group homogeneity on attitude change and extremitization in Study 5.

We utilized a variable reflecting trait openness to experience that we measured during Assignment 1 in Study 5 as the average of students' responses to four items, each of which ranged from 1 to 7 (see Chap. 2 and supplemental materials). The variable was coded so that higher values indicated higher levels of openness to experience. We interacted this variable with the variable for the homogeneity condition to predict absolute attitude change as well as extremitization during Assignments 3 and 4 in Study 5. The goal was to see if the effects of group homogeneity depended on individuals' trait levels of openness to experience.

With regard to absolute attitude change, we found a significant interaction in the expected direction but only when the homogenous groups were positive toward nanotechnology. Specifically, we found that among students who scored low in openness, there was no significant difference in absolute attitude change between students who were in a heterogeneous group or in a homogeneous group. However, among students who scored high in openness, there was significantly more attitude change in heterogeneous groups than in positive homogeneous groups. Students with high openness in negative homogeneous groups showed a statistical trend in the same direction. In other words, the expectation from the existing literature—that group heterogeneity and exposure to alternative viewpoints would lead people to alter their opinions—was only supported among students high in openness. Importantly, though, this moderation was only evident with regard to risks versus benefits item assessed during Assignment 3.

With regard to attitude extremitization, interactions are significant when predicting responses to the deregulation item for both Assignments 3 and 4, but the interactions are not significant predicting the risks versus benefits item. The pattern of the interactions corroborates the role of openness as suggested above with regard to absolute attitude change. Among students low in openness, being in the heterogeneous condition is associated with greater extremitization, but among students high in openness, being in the heterogeneous condition is associated with less extremitization than being in the homogeneous conditions (although, only significantly so in comparison to the negative homogeneous condition). Said differently, in the heterogeneous condition, openness is associated with lower levels of extremitization, whereas in the homogeneous conditions, openness is associated with greater levels of extremitization. Individuals low in openness seem most likely to polarize in the face of alternative views, but it is those who are high in openness that seem most ready to "rally the wagons" and become more extreme in their views around like-minded others. This interaction is shown in Fig. 4.4.

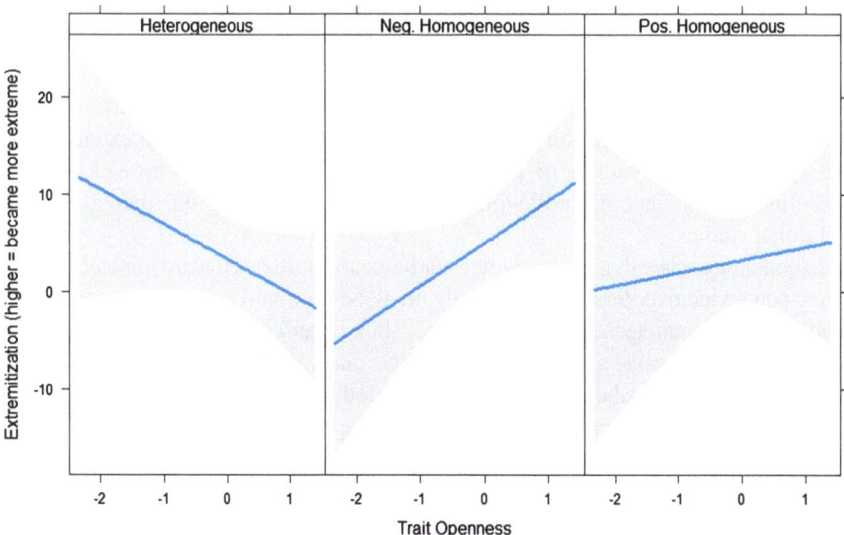

**Fig. 4.4** Interaction between group homogeneity and openness predicting attitude extremitization. Greater values on the y-axis indicate greater levels of attitude change from the prior Assignment in the direction of being more extreme in the same direction as their attitudes in the prior Assignment (positive values indicate movement to a more extreme position and negative values indicate movement in the opposite direction; in cases where one's previous attitudes were at the exact midpoint of the scale, movement in either direction is considered extremitization and thus receives positive values); the left panel reflects the heterogeneous condition, the middle panel reflects the negative homogeneous condition, and the right panel reflects the positive homogeneous condition; trait openness is represented on the x-axis

## 4.5   Conclusion: What We Have Learned and Where to Go from Here

Gauging, and at times shaping, public opinion is often a primary goal of public deliberations regarding scientific and technological development. Scientists and investors need to understand public opinion and the factors that impact it in order to know how to develop their research or technology in a publicly acceptable manner. Policymakers need to understand public opinion in order to know what regulations the public wants as well as how the public might react as development progresses. Further, researchers, investors, and policymakers may be interested in easing the fears of an apprehensive constituency or warning an overzealous public of the risks of a particular research program or technology. Finally, compared to opinion surveys that allow respondents to breeze through questions about exotic issues without serious consideration, deliberation can offer an opportunity for scientists and policymakers to understand where public opinion might go as citizens are exposed to and learn more about new, frontline technologies and research areas. Public deliberations offer an ideal setting for researchers and policymakers to interact with the public in these ways. However, as we have discussed throughout this book, the particular features of a public deliberation can have substantial implications for how participants view scientific and technological progress. In this chapter, we presented a brief smattering of results aimed at shedding light on the main effects of particular features of deliberation on attitude change and polarization. These results are far from exhaustive of the ways in which features of public deliberation might affect attitudes, and we encourage researchers and policymakers to dive deeper using our data as well as additional studies.

In general, we saw that in our studies, participants' attitudes toward nanotechnology varied somewhat over time but not greatly. In all but one study, students became more cautious about nanotechnology over time—but experienced a slight shift toward becoming more positive again by the end of the study. We did not identify any drastic attitude changes, but there were several manipulations that had notable impacts. Further, we may have missed something by averaging across the samples. Do some types of people change more than others? Are the effects of different features of deliberation universal across types of people? Indeed, we found that the various predictions from the existing literature regarding group dynamics in deliberative discussions were differentially supported depending on participants' trait openness.

The manipulations we implemented across our studies all demonstrated some level of consistency in the direction of their effects, but significant differences were sporadic and modest. The most consistent findings seemed to be that critical thinking prompts and information structured in terms of alternative perspectives moved participants toward more heavily weighing the risks of nanotechnology and to some degree led to less attitude change but in a direction of becoming less extreme when change did occur. However, even these findings were not entirely consistent throughout studies. This suggests that the features of deliberation we manipulated are indeed promising as potential subjects for further investigation, but it cannot be said that these features have large and ubiquitous effects on attitudes in the population studied. On the one hand, this means we have yet to uncover features of deliberation that consistently produce "positive" outcomes. On the other hand, it means the manipulations we implemented, which reflect commonly used features in prominent deliberations, did not result in the adverse outcomes feared by skeptics. Furthermore, as we showed with our investigation of trait openness, there are likely various mediators and moderators of the effects of the features we looked at, which can be examined using our data or in future studies.

Attitude change and polarization are issues that public deliberation organizers cannot afford to ignore. Polarization and gridlock on scientific and technological issues can put a complete halt to development, as can widespread public skepticism. Yet too much enthusiasm can lead policymakers to forgo the careful consideration necessary to form effective regulations. Public deliberations offer researchers and policymakers an opportunity to nip these potential crises in the bud. However, a scientific understanding of the effects of different features of deliberation is necessary in order to ensure that deliberation does not make things worse instead of better.

# References

Alessandro, N., Manzo, A., Veronesi, F., & Rosellini, D. (2013). An overview of the last 10 years of genetically engineered crop safety research. *Critical Reviews in Biotechnology, 34*, 1–12.

Bettencourt, B. A., & Dorr, N. (1998). Cooperative interaction and intergroup bias: Effects of numerical representation and cross-cut role assignment. *Personality and Social Psychology Bulletin, 24*(12), 1276–1293.

Bodenhausen, G. V., & Macrae, C. N. (1998). Stereotype activation and inhibition. *Stereotype activation and inhibition: Advances in social cognition, 11*, 1–52.

Bornstein, G. (1992). The free-rider problem in intergroup conflicts over step-level and continuous public goods. *Journal of Personality and Social Psychology, 62*, 597–606.

Bouas, K. S., & Komorita, S. S. (1996). Group discussion and cooperation in social dilemmas. *Personality and Social Psychological Bulletin, 22*, 1144–1150.

Cacioppo, J. T., Petty, R. E., Feinstein, J. A., & Jarvis, W. B. G. (1996). Dispositional differences in cognitive motivation: The life and times of individuals varying in need for cognition. *Psychological Bulletin, 119*(2), 197.

Chambers, S. (2003). Deliberative democratic theory. *Annual review of political science, 6*(1), 307–326.

Charles, D. (2016). *Congress just passed a GMO labeling bill. Nobody's super happy about it.* Retrieved from http://www.npr.org/sections/thesalt/2016/07/14/486060866/congress-just-passed-a-gmo-labeling-bill-nobodys-super-happy-about-it.

Cohen, J. (1998). Democracy and liberty. *Deliberative democracy, 1*, 185.

Davis, J. H., Kameda, T., Parks, C., Stasson, M., & Zimmerman, S. (1989). Some social mechanics of group decision making: The distribution of opinions, polling sequence, and implications for consensus. *Journal of Personality and Social Psychology, 57*, 1000–1012.

Delli Carpini, M. X., Cook, F. L., & Jacobs, L. R. (2004). Public deliberation, discursive participation, and citizen engagement: A review of the empirical literature. *Annual Review of Political Science, 7*, 315–344.

Dewey, C. (2017). *The government is going to counter 'misinformation' about GMO foods.* Retrieved from https://www.washingtonpost.com/news/wonk/wp/2017/05/03/the-government-is-going-to-try-to-convince-you-to-like-gmo-foods/?utm_term=.01c6dcc416d9.

Dewey, J. (1927). *The public and its problems.* Athens, OH: Swallow.

The Economist. (2014). *Vermont v Science.* Retrieved from https://www.economist.com/news/united-states/21601831-little-state-could-kneecap-biotech-industry-vermont-v-science.

Festinger, L. (1957). *A theory of cognitive dissonance.* Stanford, CA: Stanford University Press.

Fishkin, J. S. (1991). *Democracy and deliberation: New directions for democratic reform* (Vol. 217). New Haven, CT: Yale University Press.

Fishkin, J. S., Iyengar, S., & Luskin, R. C. (2005). *Deliberative public opinion in presidential primaries: Evidence from the online deliberative poll.* Paper read at International Communication Association Annual Meeting, at New York, NY.

Fishkin, J. S., & Luskin, R. C. (1999). Bringing deliberation to the democratic dialogue. In *The poll with a human face: The National Issues Convention experiment in political communication* (pp. 3–38).

Funk, C., & Rainie, L. (2015). *Public and scientists' views on science and society.* Retreived from http://www.pewinternet.org/2015/01/29/public-and-scientists-views-on-science-and-society/.

Gaertner, S. L., & Dovidio, J. F. (2014). *Reducing intergroup bias: The common ingroup identity model.* New York, NY: Psychology Press.

Gutmann, A., & Thompson, D. (1996). *Democracy and disagreement: Why moral conflict cannot be avoided in politics, and what can be done about it.* Cambridge, MA: Harvard University Press.

Gutmann, A., & Thompson, D. (2009). *Why deliberative democracy?* Princeton, NJ: Princeton University Press.

Habermas, J. (1989). *The structural transformation of the public sphere. Thomas burger* (Vol. 85, pp. 85–92). Cambridge, MA: MIT Press.

Habermas, J. (1996). *Between facts and norms*. Cambridge, MA: MIT Press. (W. Rehg, Trans.).

Insko, C. A., Schopler, J., Drigotas, S. M., Graetz, K. A., Kennedy, J., Cox, C., & Bornstein, G. (1993). The role of communication in Interindividual-intergroup discontinuity. *Journal of Conflict Resolution, 37*, 108–138.

Isenberg, D. J. (1986). Group polarization: A critical review and meta-analysis. *Journal of Personality and Social Psychology, 50*, 1141–1151.

Kahan, D. M., Peters, E., Wittlin, M., Slovic, P., Ouellette, L. L., Braman, D., & Mandel, G. (2012). The polarizing impact of science literacy and numeracy on perceived climate change risks. *Nature Climate Change, 2*(10), 732.

Kam, C. D., & Estes, B. A. (2016). Disgust sensitivity and public demand for protection. *The Journal of Politics, 78*(2), 481–496.

Karlamangla, S. (2014). *L.A. backpedals on proposal to ban growing genetically modified crops.* Retrieved from http://beta.latimes.com/local/cityhall/la-me-1209-gmo-vote-20141209-story. html.

Kingsbury, N. (2009). *Hybrid: The history and science of plant breeding*. University of Chicago Press.

Lewandowsky, S., Gignac, G. E., & Oberauer, K. (2013). The role of conspiracist ideation and worldviews in predicting rejection of science. *PloS One, 8*(10), e75637.

Maass, A., & Clark, R. D. (1984). Hidden impact of minorities: Fifteen years of minority influence research. *Psychological Bulletin, 95*(3), 428.

Marcus, G. E., Neuman, W. R., & MacKuen, M. (2000). *Affective intelligence and political judgment*. Chicago, IL: University of Chicago Press.

Marris, C., Wynne, B., Simmons, P., & Weldon, S. (2001). Public perceptions of agricultural biotechnologies in Europe. Final Report of the PABE Research Project. Commission of European Communities.

McCarty, N., Poole, K. T., & Rosenthal, H. (2006). *Polarized America: The dance of political ideology and unequal riches*. Cambridge, MA: MIT Press.

McCright, A. M., & Dunlap, R. E. (2011). The politicization of climate change and polarization in the American public's views of global warming, 2001–2010. *The Sociological Quarterly, 52*(2), 155–194.

McLean, I. S., List, C., Fishkin, J. S., & Luskin, R. C. (2000). *Does deliberation produce preference structuration? Evidence from deliberative opinion polls* (http://www.la.utexas.edu/research/delpol/papers/structuration.pdf). Paper read at American Political Science Association Annual Meeting, at Washington, DC.

Mendelberg, T. (2002). The deliberative citizen: Theory and evidence. *Political decision making, deliberation and participation, 6*(1), 151–193.

Moscovici, S. (1985). Innovation and minority influence. In G. Lindzey & E. Aronson (Eds.), *The handbook of social psychology* (Vol. 2, pp. 347–412). New York, NY: Random House.

Moscovici, S., & Mugny, G. (1983). Minority Influence. In P. B. Paulus (Ed.), *Basic Group Processes* (pp. 41–64). New York, NY: Springer Series in Social Psychology. Springer.

Muhlberger, P., Gonzalez, F. J., PytlikZillig, L. M., Hutchens, M. J., & Tomkins, A. J. (2017). *Estimating multiple forms of attitude polarization and change during deliberative discussion.* Unpublished Manuscript.

Mutz, D. C. (2006). *Hearing the other side: Deliberative versus participatory democracy.* Cambridge, MA: Cambridge University Press.

Myers, D. G., Bruggink, J. B., Kersting, R. C., & Schlosser, B. A. (1980). Does learning others' opinions change one's opinion? *Personality and Social Psychological Bulletin, 6*, 253–260.

Petty, R. E., Haugtvedt, C. P., & Smith, S. M. (1995). Elaboration as a determinant of attitude strength: Creating attitudes that are persistent, resistant, and predictive of behavior. In R. E. Petty & J. A. Krosnick (Eds.), *Attitude strength: Antecedents and consequences* (Vol. 4, pp. 93–130). New York, NY: Psychology Press.

Price, V., & Cappella, J. N. (2002). Online deliberation and its influence: The electronic dialogue project in campaign 2000. *IT & Society, 1*(1), 303–329.

Sally, D. (1995). Conversation and cooperation in social dilemmas: A meta-analysis of experiments from 1958 to 1992. *Rationality and Society, 7*, 58–92.

Schkade, D., Sunstein, C. R., & Kahneman, D. (2000). Deliberating about dollars: The severity shift. *Columbia Law Review, 100*, 1139–1175.

Schulz-Hardt, S., Frey, D., Lüthgens, C., & Moscovici, S. (2000). Biased information search in group decision making. *Journal of Personality and Social Psychology, 78*(4), 655.

Scott, S. E., Inbar, Y., & Rozin, P. (2016). Evidence for absolute moral opposition to genetically modified food in the United States. *Perspectives on Psychological Science, 11*(3), 315–324.

Smith, C. M., Tindale, R. S., & Dugoni, B. L. (1996). Minority and majority influence in freely interacting groups: Qualitative versus quantitative differences. *British Journal of Social Psychology, 35*, 137–149.

Taber, C. S., & Lodge, M. (2006). Motivated skepticism in the evaluation of political beliefs. *American Journal of Political Science, 50*(3), 755–769.

Tetlock, P. E., & Kim, J. I. (1987). Accountability and judgment processes in a personality prediction task. *Journal of Personality and Social Psychology, 52*(4), 700.

# Chapter 5
# Policy Acceptance

**Abstract** This chapter focuses on the predictors of policy acceptance and upon elucidating the pathways through which different features of public engagement might impact such acceptance—especially when a policy is not preferred. Examination of the relationships between experimentally varied features of engagement and policy acceptance suggests few, if any, main effects of different features. There is even less evidence that any of the engagement features change the relationship between policy preferences and policy acceptance. However, a more fine-grained analysis suggests a more nuanced story. There was evidence that certain features of engagement, such as promoting discussion or encouraging critical thinking, impacted mediators and moderators such as conscientious engagement and negative perceptions of the information that was provided. These mediating or moderating variables, in turn, impacted policy acceptance and/or the relationship between policy preferences and policy acceptance—sometimes in a manner that suggested competing pathways were at work, cancelling one another out, and resulting in our apparent "null effects." Our results also varied dependent upon whether the policy context was one of relative risk (promoting the development of nanotechnology) or one of relative status quo (promoting slow development and higher regulation of nanotechnology). Thus, our results suggest a fuller understanding of the impacts of engagement features on hoped-for outcomes (like policy acceptance) requires careful attention to causal pathways that operate in different policy contexts.

**Keywords** Policy acceptance · Policy preference · Policy resistance · Regulation · Process fairness · Process perceptions · Procedural justice

**Electronic supplementary material**: The online version of this chapter (https://doi.org/10.1007/978-3-319-78160-0_5) contains supplementary material, which is available to authorized users.

## 5.1   Introduction

When President Trump indicated he might not accept the results of the 2016 election, the protests were loud and numerous. The USA prides itself on peaceful transitions of power, and, for some, Trump's reluctance to say he would accept the election results if he lost was taken as a lack of loyalty and commitment to democracy. *Real* Americans, one could argue, accept the outcomes of our democratic processes, whether we like them or not, especially when it comes to transitions of power. In his defense, Trump claimed that the processes might not be democratic at all, but instead "rigged." But perhaps most interestingly, when Trump won the election, the tables turned. After the election, Trump appeared less concerned about the possibility that the wrong processes had been used, and others, likely including some others who had accused Trump of disrespecting our democracy, were now more concerned that maybe the election was rigged after all.

It seems clear that people's assessments of decision-making *processes* and *outcomes* are intertwined, with such assessments likely having bidirectional influences upon each other. The dual influences of perceived processes and outcomes have long been recognized in the literatures examining distributive justice, which focuses on the fairness of outcomes, and procedural justice, which focuses on the fairness of the procedures leading to those outcomes (Cropanzano & Folger, 1991; Lind & Tyler, 1988). Trust likely also plays a role (Brockner, 1996). Those most likely to question Trump's commitment to democracy appeared to be those who also trusted him the least. Likewise, Trump's own portrayal of the reasons why he might not accept the election results questioned the trustworthiness of the processes (are they fair or rigged?), the media (are they honest and competent or corrupt and incompetent?), and the Democratic and even Republican parties (do they care about the American public or only about their own interests?).

In a democracy such as the USA, processes are put into place with the intention and hope that the outcome will be accepted, even if (or especially if) not preferred, on the basis that appropriate processes were used. Acceptance might also depend on perceptions that the involved parties behaved trustworthily during the processes (i.e., in competent, caring, honest ways) (Mayer, Davis, & Schoorman, 1995) and can be trusted to follow and respect any implied outcomes of such processes. The outcomes are not always "accepted" entirely, but one hope that underlies our democracy is that the outcomes will be at least sufficiently accepted for a time—at least until the time that they are overturned, again, via the appropriate processes, as opposed to via violence.

The key words here are "acceptance" and "appropriate processes." Instead of having wars to decide who has power and who gets to decide what rules are in place, democracies have set up *other* processes by which to decide such things. Ideally, the best ideas will win, and people who disagree with decisions made through the established processes will nonetheless accept—or at least tolerate and abide by such decisions—at least for a time, that is, until the time such that the decision is changed by additional rounds of appropriate processes.

But what are appropriate processes? More central to the purposes of our studies, are there certain *public engagement processes* that can be used during decisions about policies, in order to increase public acceptance of decisions and the policies arising from such processes? Are there processes that might even reduce the impact of one's policy preferences on policy acceptance, so that those who do not receive their preferred outcome are nonetheless willing to accept the outcome? Meanwhile, do any public engagement processes actually make things worse? Are there processes that decrease the likelihood that people will accept policies that are not preferred? And in either case, what accounts for the linkages between various public engagement processes and policy acceptance?

In this chapter, we examine variables that predict acceptance or support of nanotechnology policies across one or more of our four studies. In so doing, we also examine whether our experimental manipulations impact some theorized mediators and moderators, such as process fairness, conscientious engagement, and trustworthiness perceptions. Furthermore, given the importance of accepting polices even when they are not preferred, we explore factors that might alter the relationship between "policy preferences" and "policy acceptance," by increasing or decreasing their dependence upon each other. As we give an overview of our results, we point to some findings that seem predictable and understandable. Other findings, however, may leave the reader (as well as we writers) uncomfortably searching for explanations. We hope this spurs more research and not just frustration. But before we get to that, let us first provide some theoretical context.

## 5.2 A Rough Draft Theory of Policy Preference, Acceptance, and Support

### 5.2.1 For What? Definitions and Relationships Between Some Key Variables

The outcome of greatest interest to this chapter is policy acceptance and support. While it may not always be important to make such fine-grained distinctions, because of our interest in whether certain public engagement processes might increase the *acceptance* or *support* of policies even when they are not *preferred,* we need to distinguish between such constructs. Currently, there is a noted lack of existing theory distinguishing and connecting policy preferences, acceptance, and support (Dreyer, Polis, & Jenkins, 2017). We draw from the little that we could find (e.g., see also Rau, Schweizer-Ries, & Hildebrandt, 2012), as well as literatures that we perceive might be related (i.e., the literatures relating to legitimacy and procedural fairness) to develop our theoretical approach.

Our view of the relationships between constructs is illustrated in Fig. 5.1, which shows policy preference ranging from low to high along the horizontal dimension. We define *policy preference* as an evaluative attitude of preferring, liking, agreeing

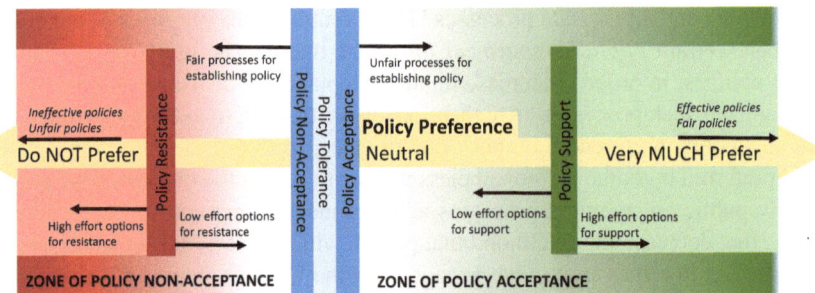

**Fig. 5.1** An illustration of the relationships conceptualized among policy preference attitudes (horizontal light-colored yellow line), acceptance/nonacceptance (double vertical blue border and zones to right and left of each), and support/resistance (subzones within the red nonacceptance and green acceptance zones, which begin at the point of the single vertical borders and extend to the extremes)

with, and being "for" a given policy. Our construct of preference is similar to constructs suggested by Schade and Schlag (2003) and Dreyer and colleagues' (Dreyer & Walker, 2013). Notably, these authors used terms like acceptance and acceptability for their similar constructs—whereas we argue that "preference" is a better term and that acceptance should be viewed as distinct from preference.

We use the term *policy acceptance* to refer to judgments and evaluations about the policy *being in place*. As we asked our participants in Studies 2–4, do you agree that the government made the right decision [by adopting the policy]? Acceptance is expected to be strongly affected by policy preferences but also affected by other factors such as processes used to make the policy. One could, for example, express acceptance of a policy (or some political outcome, like a presidency, to return to our earlier example) that one does not prefer. This could happen perhaps due to an assessment that the policy decision (or decision about who is in power) was arrived at fairly.

Although it is less important to our current analyses to distinguish between policy acceptance and support, we do so because it is important theoretically, and in case later it helps to make sense of some of our more puzzling results. Prior authors have distinguished policy support as reflecting behaviors, while constructs like policy preference or acceptance reflect attitudes (Batel, Devine-Wright, & Tangeland, 2013; Dreyer, Teisl, & McCoy, 2015). Policy support (and its opposite, policy resistance) is viewed as a behavioral expression of attitudes toward the policy. Therefore, models of policy acceptance and support generally view policy acceptance and nonacceptance attitudes as perquisite to policy support and resistance behaviors.

Regarding the relationships between these constructs, as shown in Fig. 5.1, we theorize that policy acceptance, nonacceptance, and policy support and resistance usually occur in different zones of the policy preference dimension. Acceptance and support are expected to be more likely at higher levels of policy preference. Nonacceptance and resistance are expected to occur at lower levels of policy preference. And like Dreyer and colleagues, we theorize that the boundary between acceptance and nonacceptance will occur at more moderate levels of policy preference (i.e., closer to neutral) than active support and resistance.

## 5.2.2   *What Works and How? Prior Research and Theory Concerning Factors Impacting Policy Acceptance and Support*

In the past, and usually without carefully distinguishing between the constructs we just reviewed, a number of factors have been found to impact policy preference, acceptance, and support across a range of policy types. Although we do not test all of these relationships in the analyses presented in this chapter, the small black arrows shown in Fig. 5.1 reflect our current educated guesses as to examples of certain variables that may act more directly on each construct as defined here. For example, perceived fairness and effectiveness of a policy are commonly found to predict attitudes toward and willingness to support a policy (Drews & van den Bergh, 2016; Dreyer & Walker, 2013). Because the perceptions are appraisals directly relevant to the policies themselves, it is likely they most strongly correlate with and affect preferences for those policies.

As another example, Dreyer et al. (2015) note that whether or not people behave in a supportive way will be influenced by how much effort a behavior is required. They argue it takes less effort to form an attitude than to act upon it and that policy acceptance is usually necessary for policy support (Dreyer et al., 2017). Consistent with their thinking about effort, recent political mobilization efforts use technology to change the amount of effort required to engage in support/resistance (e.g., sending a letter to one's representative) by creating pre-written letters and forms already addressed to the appropriate representative that can be sent in less than a minute with just a few clicks of buttons on web-based forms.

Especially relevant to our studies, deliberative and participatory engagement has been hypothesized to increase policy acceptance via, potentially, two specific mechanisms. First, the activities might inform people (as was supported by the knowledge increases noted in Chap. 3), and new information might change people's attitudes (as was supported by attitude changes at the individual level, reported in Chap. 4). End attitudes toward topics, in turn, tend to correlate with people's policy preferences about those topics in predictable ways. In our data, for example, simple correlations between the extents to which perceived benefits outweighed the risks of nanotechnology, and to which they preferred the policy that speeded up nanotechnology development or did not prefer the policy that slowed it down, ranged from about 0.45 to 0.55. *If* people's attitudes are changed in such a way that they come to an attitudinal consensus, then most people may support the eventually chosen policy (Drews & van den Bergh, 2016). Unfortunately, this last linkage is not found in our data; recall that our results in Chap. 4 did not reveal that any of our methods resulted in a great deal of consensus that might in turn result in high policy acceptance in one experimental condition or another.

A second mechanism is that effective participatory engagement might increase positive perceptions of how the policy decisions were made. These positive perceptions might increase the perceived legitimacy of the policies themselves, which are then viewed as more acceptable to people. Thus, the boundary between acceptance and nonacceptance (in Fig. 5.1 this is represented by vertical double bars, in between which is policy tolerance) might not happen at the neutral point of one's preferences and might be found at a higher or lower point or even found to shift along the dimension of preferences, due to other factors. Consistent with this thinking, Wallner (2008) describes the role of policy legitimacy in policy failure. Waller argues that, beyond policy characteristics (such as ineffectiveness, inefficiency, and poor performance), policies can fail because the public or affected stakeholders do not find them legitimate.

In addition to having a main effect on policy acceptance, positive process perceptions might also operate through another mechanism: by reducing the effect of other variables that otherwise would reduce acceptance or support—such as negative policy preferences. Evidence for such interactions (e.g., process perception x policy preference interaction effects upon policy acceptance/support) comes from research on procedural and distributive justice (e.g., Tyler & Huo, 2002). To our knowledge, prior research has not examined such interactions in the context of studies of policy acceptance. However, quite a lot of research related to procedural and distributive justice seems relevant and may generalize. Such studies examine when and why, for example, people are willing to accept outcomes that they do not favor, such as losing in an arbitration or less-than adequate severance pay in an employment layoff. Brockner (1996) provided a list of 20 studies that found the relationship between outcome preferences and expressions of acceptance (which included things like expressions of satisfaction and organizational commitment) was moderated (i.e., reduced) by introducing decision-making and implementation procedures that were perceived as fair, respectful, transparent, inclusive, and so on. Brockner (1996) argues that the reason why such procedural justice factors have the moderating impact of reducing the influence of preferences on acceptance is because they produce trust in the persons making the decisions. In the context of our study, we might see this relationship if our measures of procedural perceptions predicted trust in policymakers which in turn moderated the relationship between policy preference and acceptance.

On the negative side, however, some have hypothesized that public engagement might *reduce* policy acceptance rather than increase it. The arguments for the potential negative impacts of public engagement on acceptance and support are somewhat similar to those discussed in Chap. 4. If, during the engagements, people bolster their belief in and the evidence for their opinions, the effect may be greater resistance to policies among those who disagree with the policy, than if they had not engaged. Indeed, prior research suggests that people who put more effort into their decisions—something that deliberative engagements seek to promote—can create more coherent attitudes which include feeling more favorable toward their choice and less favorable to alternatives (Anderson, DeTour, & PytlikZillig, 2015; Goodwin & Wu, 1984; Svenson & Jakobsson, 2010). Thus, by increasing people's certainty about their attitudes and preferences, it may be that deliberations simply make it easier for people who are for the policy to accept it while making it harder for those who are against the policy to accept it. In other words, such processes might *increase* the influence of preferences on policy acceptance/support by making those preferences more coherent.

   To summarize, and to make explicit the connections between our deliberative features and prior literature and theory, we are hypothesizing two competing processes that may affect policy acceptance/support in our studies. These processes are conceptually illustrated in Fig. 5.2. One process is hypothesized to operate through *perceptions of the processes* as being fair and competent (see Fig. 5.2a). Public engagement features that increase these perceptions would thus also increase policy acceptance/support and may also decrease the relationship between policy preferences and acceptance/support. For example, each of our manipulations (peer discussion, active facilitation, strong information, and so on) might be viewed as more fair or competent and thus increase positive perceptions of the deliberative processes. In addition, factors that affect people's perceptions of the background information—critical thinking prompts, use of strong versus weak information, and use of NIF-formatted materials—may also impact information evaluations, which we would expect to influence process perceptions, because the information was part of the process. Of course, given that process perceptions were assessed after participants learned about the policy decision outcome, we would expect that their preferences for or against the policy would impact their perceptions of processes too. If process perceptions are impacted positively by these various factors, we would expect those positive perceptions then could directly increase acceptance/support of the policies and/or indirectly increase acceptance/support by increasing trust in policymakers, as well as by reducing the relationship between preferences and acceptance/support (moderation effect shown with a dotted line).

   The second process (see Fig. 5.2b) is hypothesized to operate through *creating stronger and more coherent attitudes* toward the topics (nanotechnology) and policies under discussion. For example, prompting people to think in an effortful and conscientious manner might not only increase knowledge as suggested by Chap. 3 findings but also increase how certain people are about their resulting attitudes. Use of peer discussion and active facilitation might also increase such deliberative efforts. Use of strong information that is optimally formatted (e.g., using NIF organization) might improve people's perceptions of the information upon which their decisions are based and thus further increase their confidence in their attitudes. Confidence in one's attitudes would then increase the strength of the relationship between one's preferences and policy acceptance/support. However, because we did not directly assess people's certainty in their attitudes, we could only look at the potential moderating impacts of deliberative engagement and information evaluations instead (moderating effects shown with dotted lines).

   Importantly, these two competing processes might involve the same public engagement features. For example, strong information might increase positive process perceptions *and* increase attitude certainty, resulting in both positive and negative effects on acceptance. Such mixed effects could hide any main effects of strong information on policy acceptance, making it important to examine the effects of our experimental manipulations, not only on the desired end but on the mediators by which they are thought to work.

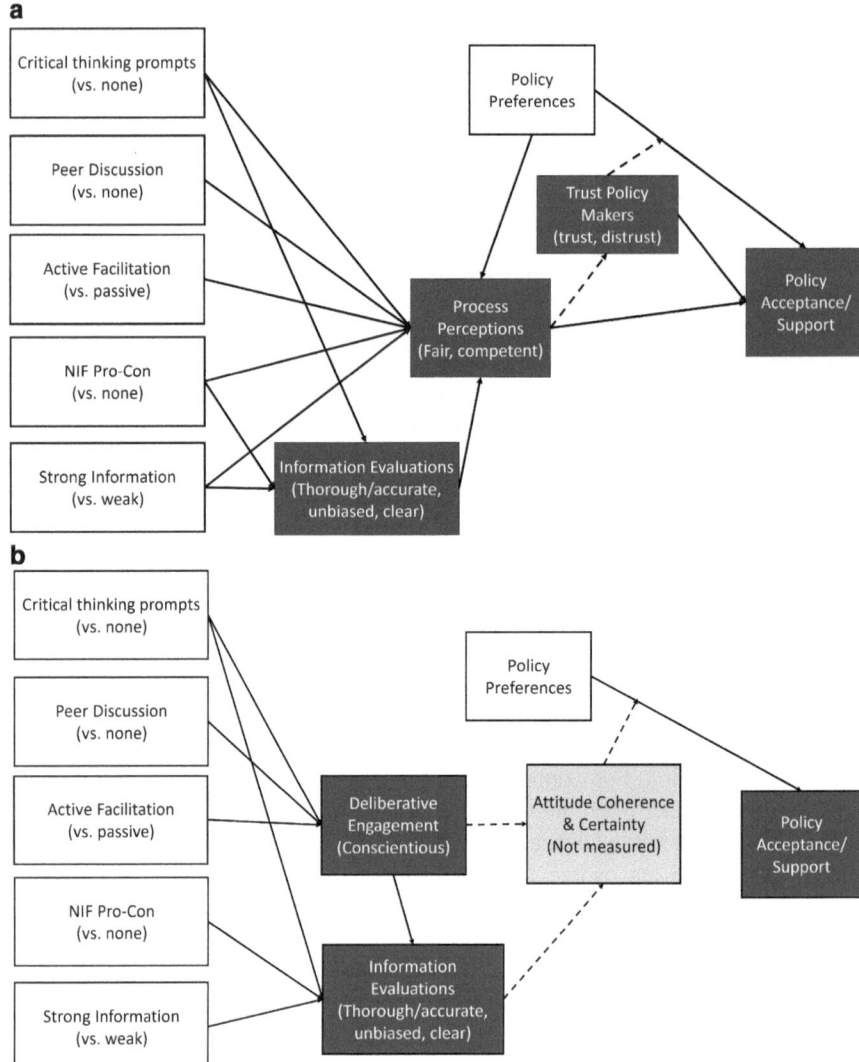

**Fig. 5.2** Theoretical models relating our experimental manipulations and mediator/moderator variables to policy acceptance/support. (**a**) Process perceptions model. (**b**) Attitude coherence model

## 5.3   The Current Study

The overarching research question we focus on in context of nanotechnology policy and the present analyses is: *What factors—especially factors relating to features of public engagement—predict willingness to accept/support a policy decision, whether or not that policy decision is/was preferred?* Because the variables available across studies varied, for the purposes of this chapter, we did not do formal path analyses to explore the relationships depicted in Fig. 5.2. Instead, using simple correlations and multiple regression analyses, we examined, across studies, three categories of results: (1) whether our experimental manipulations, overall, appear to impact policy acceptance/support and/or the degree to which policy preference relates to policy acceptance/support, (2) whether our experimental manipulations impacted various potential mediators that might have effects on policy acceptance/support, and (3) whether our potential mediators did indeed predict policy acceptance/support and/or moderate the impact of policy preference on policy acceptance.

### 5.3.1   The Policy Scenarios

We were especially interested in how participants viewed the use of public input in their decision-making processes leading up to the policy choice. Thus, as noted in Chap. 2, our method for measuring policy preferences and acceptance also purposely increased the salience of the public engagement processes used just prior to asking questions tapping policy acceptance/support and policy preferences. Specifically, in each of our studies, after all of the engagement activities, a scenario asked the participants to imagine that the very processes to which they had been exposed (which, of course, varied by experimental condition) were used to gather public input and that public input was then used to make a policy decision. In the scenario, the government's decision—to either invest more in nanotechnology development and decrease regulations (pro-development of nanotechnology) or invest less in development and increase regulations (pro-regulation of nanotechnology)—was always portrayed to be consistent with the public input obtained from the engagement activities. Furthermore, the scenario was randomly assigned so that participants had roughly equal chances of receiving a scenario that did or did not match their preferences.

### 5.3.2   Key Variables

While many details of our measures and manipulations are given in Chap. 2 and in our detailed methodological reports, here we review those variables most important to this chapter. Following the scenario, we assessed our dependent variable, which we call *policy acceptance/support*, because we treat acceptance and support the same (without distinction) across our models and in our results. However, it may be important to note that acceptance/support was assessed in a manner more closely matching the conceptualization of policy acceptance in Studies 2–4, where we asked participants whether they agreed the government made the right choice and if they agreed that the government made the same choice they would have made. In Study 5, our acceptance/support variable more closely reflected policy support, as we directly asked about willingness to support versus resist the policies as well as whether they accepted the decision due to the processes used to come to the decision.

Immediately after the scenario, we also asked about perceptions of the processes used to make the policy decision, including how fair and competent[1] the process was. The scenarios stressed that the public input methods used in the scenario were the same ones used with the survey respondent. Thus, we expected different process perceptions might occur as a result of students having been in different experimental conditions.

Finally, in each of our studies, our measure of *policy preference* was comprised of a single item. In Studies 2–4, separate from but immediately after the scenario was administered and a few questions about the scenario were answered, participants were randomly assigned to one of two versions of the policy preference item: "If legislation were being considered that would slow down (speed up) nanogenomics research and development in the area of human enhancement by decreasing (increasing) funding and increasing (decreasing) restrictions...Would you be FOR or AGAINST such legislation?" Responses fell on a six-point scale ranging from "strongly AGAINST" to "strongly FOR" with no neutral point. Items were recoded to reflect whether or not the participants preferred the policy randomly assigned to them in the scenario.[2] Thus "pro-slow" development of nanotechnology persons were identified as those who were for slowing down the research and development (or against speeding it up) and "pro-fast" if they were for speeding it up (or against slowing it down). In Study 5, the policy preference item was not separate from the scenario but instead referenced the scenario and read: "If the legislation above [in the scenario] were really under consideration by the government...Would you be FOR or AGAINST such legislation?" It was followed by the same six-point scale used in Studies 2–4.[3]

Other variables in our model were also assessed. We used participants' average reported "conscientious" engagement across all of the times measured, to

---

[1] In Studies 2–4 perceived process competence was estimated by asking participants if they felt the government should have used a different process. A factor analysis suggested this item fit with the competence scale in Study 5 along with the item asking if they felt the process was incompetent.

[2] Although these are not equivalent questions, we use them as rough approximations for preferences. The use of these two questions was designed to allow for an exploration of question framing effects which we do not delve into here.

[3] This, of course, differs from the prior studies in that it likely creates some additional carryover between the scenario and the respondents' expression of policy preferences.

operationalize effortful, deliberative engagement. We used this variable because, as noted in Chap. 2, it was most reliably related to participants feeling as though they had gained knowledge from the activities. The conscientious engagement measure was available for all four studies. We used three negative valence scales assessing participant perceptions of the background materials as biased, unclear, and untrustworthy (i.e., not accurate, not thorough). These three scales were available in Studies 3, 4, and 5; Study 2 had only two of the scales (biased and untrustworthy scales). Trust in policymakers was variably assessed across studies, with only Studies 3 and 4 providing measures of perceived trustworthiness and perceived untrustworthiness of policymakers across all participants. In Study 3, we additionally measured perceptions of policymaker's ability and motivation to take into account ethical, legal, and social issues when considering the policies that they were making.

## 5.4  Analyses and Results

As previously noted, we broke our broad research question about "what works and how" into three smaller questions: (1) Do our experimental manipulations, overall, impact policy acceptance/support and/or the degree to which policy preference relates to policy acceptance/support? Although little evidence was found for the main or moderating impacts of our experimental manipulations on acceptance/support, given that we had theorized competing processes might be in play, we next conducted analyses to explore: (2) Do our experimental manipulations impact the various potential mediators illustrated in Fig. 5.2 which may have effects on policy acceptance/support? Finally, our analyses explored: (3) Do the mediators predict policy acceptance/support and/or moderate the impact of policy preference on policy acceptance? To answer these questions, we both examined the correlations among our variables and results from multiple regression analyses.

### 5.4.1  Simple Correlations

Table 5.1 shows the simple correlations between each of our dummy-coded experimental conditions and the other variables in the study. Table 5.2 displays the correlations between other major measured variables, showing the individual correlations for each relevant study below the diagonal and the average of those simple correlations above the diagonal.

In Table 5.2, the largest average correlations occurred as might be expected: (1) among similar sets of variables (e.g., the average correlations involving the perceptions of information ranged $|r| = 0.38–0.56$; the ones involving perceptions of process fairness and incompetence ranged from $-0.41$ to $-0.55$ for an average of $-0.48$ across studies), (2) between policy preference and policy acceptance (average $r$ across studies = 0.54), and (3) among perceptions of the process (fairness

**Table 5.1** Correlations between experimental conditions (variables 1–5) and other model variables

| | Scen. | Policy prefer. | Process perceptions | | Information perceptions | | | Policymaker trustworthiness | | | Consc. eng. | Policy accept |
|---|---|---|---|---|---|---|---|---|---|---|---|---|
| | 6 | 7 | 8 | 9 | 10 | 11 | 12 | 13 | 14 | 15 | 16 | 17 |
| 1. Critical thinking = 1 | .09 | .01 | -.04 | .16* | .11 | .16+ | .14* | -.08 | .01 | -.05 | -.16* | -.05 |
| | .01 | -.02 | -.01 | .15* | .35*** | .26*** | .30*** | .03 | -.10 | | .10+ | -.05 |
| | .02 | -.02 | .07 | -.13+ | .09 | .22** | .09 | | | | .03 | .05 |
| | -.02 | .05 | -.06 | .01 | .22*** | .14* | .09 | | | | .03 | .01 |
| 2. NIF condition = 1 (S2-4) | -.04 | -.02 | -.06 | -.11 | -.01 | -.03 | -.10 | -.01 | .09 | .00 | .01 | -.02 |
| | -.04 | .00 | .07 | -.10 | -.15** | -.10+ | -.02 | -.10 | .14* | | -.03 | .02 |
| | -.10 | -.03 | -.10 | .14+ | .02 | .11 | .07 | | | | -.10 | -.10 |
| 3. Strong information = 1 (S5) | -.03 | -.04 | .11 | .02 | -.06 | .05 | -.04 | | | | .12* | -.01 |
| 4. Peer discussion = 1 (S2-4) | .04 | .04 | .01 | .00 | .03 | .01 | -.06 | .09 | -.03 | .09 | .18* | -.02 |
| | -.04 | -.06 | -.04 | .11+ | .04 | .06 | -.02 | .06 | -.01 | | .15* | .00 |
| | .05 | -.04 | -.07 | .06 | -.05 | .04 | .06 | | | | .16* | -.04 |
| 5. Active facilitation = 1 (S5) | -.09 | .12+ | .04 | -.11+ | -.08 | -.10+ | -.13* | | | | .07 | .14* |

Note: Variables 1–6 were dummy-coded variables representing our experimental conditions, with 1 denoting the indicated condition. Correlations within each row are for each study in order (2, 3, 4, 5) when available. When variables were not available for all Studies 2–5, the studies they were available for are indicated.

Abbreviations: *Scen.* scenario condition reflecting pro-development (1) or pro-regulation (0) of nanotechnology, *Policy prefer.* policy preference, *Consc. eng.* conscientious engagement, *Policy accept* policy acceptance

Numbered variables are the same as in Table 5.2: 8 = process fair, 9 = process incompetent, 10 = information biased, 11 = information untrustworthy, 12 = information not clear, 13 = policymakers trustworthy, 14 = policymakers untrustworthy, 15 = policymakers ELSI-specific trustworthiness

+p < .10
*p < .05
**p < .01
***p < .001

**Table 5.2** Correlations among major study variables by study (below diagonal) and average correlations across studies (above diagonal)

| | Scen. | Policy prefer. | Process perceptions | | Information perceptions | | | Policymaker trustworthiness | | | Consc. eng. | Policy accept |
|---|---|---|---|---|---|---|---|---|---|---|---|---|
| | 6 | 7 | 8 | 9 | 10 | 11 | 12 | 13 | 14 | 15 | 16 | 17 |
| 6. Scenario (1 = pro-devel.) | — | -.04 | -.08 | .05 | .01 | .05 | .00 | -.03 | -.05 | -.10 | -.01 | -.17 |
| 7. Policy preference | .00<br>-.02<br>-.09<br>-.07 | — | .35 | -.30 | .04 | -.01 | -.04 | .01 | .02 | -.01 | .01 | .54 |
| 8. Process fair | -.06<br>-.08<br>-.07<br>-.11 | .50***<br>.34***<br>.21**<br>.36*** | — | -.48 | -.15 | -.12 | -.11 | .03 | .04 | .07 | .06 | .59 |
| 9. Process incompetent | .08<br>-.01<br>.05<br>.09 | -.38***<br>-.22**<br>-.21**<br>-.39*** | -.55***<br>-.43***<br>-.41***<br>-.52*** | — | .14 | .10 | .10 | -.01 | .02 | .03 | -.02 | -.50 |
| 10. Information biased | -.04<br>-.01<br>.01<br>.06 | .00<br>.10+<br>-.02<br>.07 | -.25**<br>-.02<br>-.19**<br>-.15* | .12<br>.14*<br>.16*<br>.14* | — | .56 | .38 | -.13 | .06 | -.20 | -.15 | -.07 |
| 11. Information untrustworthy | .04<br>.02<br>.07<br>.05 | -.06<br>.06<br>-.02<br>-.01 | -.07<br>-.04<br>-.21**<br>-.17* | .11<br>.06<br>.06<br>.17** | .51***<br>.59***<br>.55***<br>.58*** | — | .49 | -.24 | .13 | -.28 | -.27 | -.06 |
| 12. Information not clear (S3-5) | .03<br>-.02<br>-.02 | -.09<br>.02<br>-.03 | -.06<br>-.09<br>-.17* | .05<br>.04<br>.21** | .30***<br>.42***<br>.41*** | .34***<br>.62***<br>.51*** | — | -.08 | .06 | -.14 | -.31 | -.08 |
| 13. Policymakers trustworthy (S3-4) | .00<br>-.07 | -.01<br>.03 | .02<br>.05 | .07<br>-.10 | -.16*<br>-.10 | -.26***<br>-.22** | -.06<br>-.10 | — | -.42 | .68 | .19 | .15 |
| 14. Policymakers Distrustworthy (S3-4) | -.10<br>-.01 | -.01<br>.05 | .10<br>-.02 | -.04<br>.07 | .07<br>.06 | .07<br>.19** | .01<br>.10 | -.40***<br>-.45*** | — | -.35 | -.10 | -.05 |

(continued)

**Table 5.2** (continued)

| | Scen. | Policy prefer. | Process perceptions | | Information perceptions | | | Policymaker trustworthiness | | | Consc. eng. | Policy accept |
|---|---|---|---|---|---|---|---|---|---|---|---|---|
| | 6 | 7 | 8 | 9 | 10 | 11 | 12 | 13 | 14 | 15 | 16 | 17 |
| 15. ELSI-specific trustworthiness (S3) | -.10+ | -.01 | .07 | .03 | -.20** | -.28*** | -.14* | .68*** | -.35*** | -- | .20 | .18 |
| 16. Conscientious engagement | .00<br>-.11+<br>.05<br>.03 | .10<br>.05<br>-.10<br>-.03 | .14+<br>.05<br>-.07<br>.12+ | .03<br>-.00<br>-.01<br>-.10 | -.18*<br>-.07<br>-.11<br>-.25*** | -.24**<br>-.18**<br>-.32***<br>-.33*** | -.31***<br>-.32***<br>-.29*** | .12+<br>.26*** | .00<br>-.20** | .20** | -- | .02 |
| 17. Policy acceptance/ support | -.13+<br>-.11+<br>-.29***<br>-.16* | .56***<br>.49***<br>.33***<br>.78*** | .71***<br>.54***<br>.57***<br>.53*** | -.57***<br>-.33***<br>-.49***<br>-.59*** | -.08<br>.00<br>-.17*<br>-.01 | .00<br>.01<br>-.17<br>-.06 | -.12+<br>-.04<br>-.08 | .07<br>.23** | -.02<br>-.09 | .18** | .01<br>.06<br>-.05<br>.05 | -- |

Note: Variable 6 is dummy coded with 1 denoting the indicated condition. When variables were not available for all Studies 2–5, the studies they were available for are indicated. Correlations below the diagonal are for each study in order (2, 3, 4, 5) when available. Correlations above the diagonal are the average correlation across studies. Boxes around correlations denote correlations between variables of a similar type (e.g., process perception variables)

+p < .10
*p < .05
**p < .01
***p < .001

and competence) and policy acceptance (average |r|s = 0.50–0.59). There were also moderately strong relationships between policy preferences and process perceptions (average |r|s > 0.30). This was expected because process perceptions were measured after people learned about the policy outcome (which had been randomly assigned). This means those who received outcomes matching their preferences rated the process as more fair and competent than those who received outcomes not matching their preferences.

Other correlations are also interesting to note. Consistent with Fig. 5.2, perceptions of the information modestly predict process perceptions. Perceptions of information bias (see Table 5.2, variable 10) had the strongest relationships with perceptions of the process as fair and competent (variables 8, 9) (average |r|s = 0.14–0.15). Perceived trustworthiness of policymakers also sometimes positively correlated with policy acceptance as expected, although the correlations were not consistently statistically significant.

Diverging from the predictions illustrated in Fig. 5.2a, process perceptions did not correlate with perceptions of the trustworthiness of policymakers (average rs < 0.08, and none of the correlations for any of the studies were statistically significant). In part, this may be because participants were asked to rate policymakers in general, not the policymakers in the scenario. However, and somewhat unexpectedly, negative perceptions of the background documents (assessed during A2) were related to negative perceptions of policymakers (measured at A4). In addition, although hypothesized only as a potential moderator (see Fig. 5.2b), deliberative conscientious engagement (assessed across multiple assignments) was moderately positively related to perceptions of the information and the policymakers. That is, more conscientious engagement related to more positive perceptions of the information and policymakers. In Studies 2 and 5, conscientious engagement also was marginally predictive of process fairness perceptions.

Finally, although not the focus of our analyses, Table 5.2 shows the randomly assigned scenario (0 = pro-regulation, 1 = pro-development, variable 6) correlated with policy acceptance/support. Generally speaking, participants were more accepting and supportive of the policy scenario randomly assigned to them if it was pro-regulation rather than pro-development (rs = −0.11 to −0.29, with an average − 0.17 correlation across studies).[4]

---

[4] We also used multiple regression analyses to test the effect of scenario type after accounting for policy preferences and, if relevant, the interaction effect between preference and scenario type. The interaction effect was only significant in Study 4. In Study 4, the effect of policy preference on acceptance was stronger in the pro-development scenario and nonsignificant in the pro-regulation scenario. In Studies 2 and 3, the interaction effect was not significant, but did show the same pattern, and in Study 3 the interaction (with the same pattern) sometimes became marginally significant when other variables were controlled in our regression analyses. This suggests that, consistent with prior research on risks versus benefits, people in favor of nanotechnology development were more willing to forego benefits from nanotechnology (when they received the pro-regulation scenario) than those against nanotechnology were to embrace the potential risks (when they received the pro-development scenario).

We discuss other simple correlations as we discuss our three research questions and main categories of results, which include consideration of multiple regression analyses as well.

## 5.4.2  (1) Do Our Experimental Manipulations Impact Policy Acceptance/Support or Moderate the Policy Preference-Acceptance/Support Relationship?

The simple main effects of each of our experimental conditions were almost always not significant for the full sample. One rare simple main effect found was in Study 5, with active facilitation resulting in slightly greater policy acceptance (Table 5.1, $r = 0.14$, $p < 0.05$).

Using multiple regression procedures, we also examined the combined and interactive impacts of our manipulations on policy acceptance/support. Because our analyses so rarely revealed significant effects, we do not present the results in tables or figures. Only two statistically significant effects were found, both in Study 5. One was significant effect of group type, such that those in the negative homogenous groups were most accepting/supportive of whatever policies they received and those in the heterogeneous attitude groups were least accepting/supportive of the policies they received. Additional analyses and studies are needed to understand this result. It is possible of course that it is a chance finding. The second effect (also found in Study 5) was a significant facilitation by information interaction such that the active facilitator had a significantly positive impact on policy acceptance/support in the weak information condition only and the strong information had a significantly positive impact only when the facilitator was passive. Thus, it seemed that active facilitation compensated for the negative effects of weak information on policy acceptance/support and vice versa.

Relating to the second part of our research question, we tested for but found no significant interactions indicating our experimental manipulations had moderating effects on the relationship between policy preference and acceptance/support. That is, none of our experimental conditions appeared to significantly increase or decrease the extent to which policy preferences impacted policy acceptance.

## 5.4.3  (2) Do Our Experimental Manipulations Impact Potential Mediators?

Although our experimental manipulations rarely affected policy acceptance/support, we next explored the possibility that the experimental manipulations might have indirect effects on policy acceptance/support via impacts on other related factors, such as the process perception variables (e.g., process fairness and competence) which as Table 5.2 shows are so strongly related to policy acceptance/support. Alternatively, our manipulations might impact information evaluations (which were also related to perceptions of policymakers) or impact conscientious engagement—each of which we had hypothesized might moderate the relationship between policy preferences and acceptance/support. We also tested for potential experimental manipulation impacts on perceptions of policymaker trustworthiness perceptions.

For these analyses, we examined each potential mediator individually as a dependent variable in separate analyses, for each study. We tested for the main and interactive effects of the relevant experimental manipulations in each study, always including all main effects simultaneously, but dropped any higher-order interactions that could be ruled out by not achieving at least marginal significance ($p < 0.10$). Also, when examining effects, we usually controlled for the effect of policy preference and, when relevant, its interaction with the randomly assigned policy scenario and the main effect of the scenario. This is because, except for in the case of information perceptions (which were measured during A2 after the first reading of the information) and conscientious engagement (which was measured throughout and averaged across activities), the other variables were assessed *after* people learned about the policy outcome and may have been affected by whether they received their preferred outcome. Below we describe our main findings regarding the combined impacts of our experimental manipulations on the proposed mediators.[5]

**Process Perceptions**

Examination of the correlations in Table 5.1 revealed the experimental manipulations did not have strong or reliable relationships with process perceptions. Multiple regression analyses regressing process perceptions on our experimental manipulations simultaneously also indicated some effects, but not consistent ones. Perceptions of process fairness were only impacted by our manipulations in Study 5, such that strong information (compared to weak information) conditions were associated with higher ratings of process fairness while controlling for other manipulations. Meanwhile, ratings of process competence appeared to be affected (at least marginally) by the critical thinking conditions in the majority of our studies (again, once other manipulation effects were controlled), but not in a consistent direction. That is, (and these results are consistent with the correlation results in Table 5.1), in Studies 2 and 3, there was a main effect such that critical thinking participants felt processes were less competent than control participants. Later we note that our critical thinkers, in general, seemed critical—and this finding is consistent with that trend. However, in Study 4 the effect of critical thinking was in the opposite direction. Likewise, in Study 3 use of NIF materials increased perceived competence (i.e., decreased incompetence perceptions), but in Study 4 use of NIF materials decreased competence perceptions (increased incompetence perceptions).

---

[5] Due to space constraints, we do not present full results. However, syntax to generate full sets of the results is available from this book's first author.

Study 5 results might shed some light on the contradictory effects of critical thinking because Study 5 data suggests the effect of critical thinking depends on other factors we varied (and which could have varied unintentionally between other studies). For example, being in a homogenous positive group resulted in a significantly more positive effect of critical thinking on judgments of process competence compared to being in a homogenous negative group, in which case the effect was in the opposite (i.e., negative) direction. The effects of the critical thinking prompts also interacted with passive/active moderation and strength of information. The pattern of the three-way interaction suggested that using one positive factor (critical thinking prompts, strong information, or active moderation) could modestly increase perceived competence of the processes relative to having none of those factors. Adding a second one did not help much at all; however, you could further increase the perceived competence if all three were used. While further analyses are necessary to understand the contradictory effects found between studies, it is possible that our groups in Studies 3 and 4 were more or less homogenous in composition. It is also possible that some of our moderators were more passive than intended. Either of these situations might have impacted the direction of the critical thinking effect in Studies 3 and 4.

**Information Perceptions**
In each of our studies we found consistent evidence that critical thinking prompts impacted information perceptions. Specifically, people prompted to think critically also rated the background documents more negatively on one or more of our scales. Such effects are apparent in the correlation results shown in Table 5.1, as well as in our multiple regression results. For example, our regression results found, in Study 2, people in the critical thinking condition rated the background information as less accurate and thorough than those in the control condition. In Study 3, those in the critical thinking condition felt the information was more biased, less thorough/accurate, and less clear than those in the non-critical thinking condition.[6] In Study 4, critical thinkers again rated the background documents as less clear and less accurate/thorough.

In Study 5, partly because we had begun to wonder if our critical thinkers were just, well, critical, we created strong and weak information conditions. We wanted to see if our participants were actually sensitized to differences in quality of information rather than just rating everything about the background information more negative overall. We did, once again, find more negative evaluations of our materials among the critical thinkers. And consistent with Table 5.1 results, use of strong rather than weak information did very little to change participants' ratings of the information overall. However, in the case of ratings of bias in the background documents, there was—if we reduced our sample to those student participants who completed our post measures at A4 as well as our A2 measures—a statistically significant interaction between critical thinking prompts and information quality such that the critical thinkers were less negative about the strong information than the weak information. This gave us hope that we had actually induced critical thinking and not just negativity. Still, the fact that we needed to reduce the sample

---

[6] In Study 3 there were also marginal positive main effects of the NIF formatting of the materials on accurate/thorough and significant positive impact of NIF formatting on perceiving the documents as unbiased. However, these effects were not replicated in Study 4.

to the more "participatory" or "engaged" students in the course (i.e., those who completed A2 *and* A4 measures) suggests participant individual differences need to be taken into account in studies of public engagement, to fully understand effects that are found or not found.[7]

**Policymaker Trustworthiness**

Policymaker trustworthiness was only measured across all participants in Studies 3 and 4. The simple correlation results in Table 5.1 suggest very little impact of our manipulations on perceptions of policymaker trustworthiness or untrustworthiness. Only the use of NIF information in Study 4 appeared to be related, specifically, to increased distrust in policymakers. Multiple regression analyses further indicated that, in Study 4, the NIF-formatted materials were associated with increased perceptions of untrustworthiness overall and also related to lower perceptions of trustworthiness when participants were in the critical thinking condition. However, in Study 3, a different effect was found: There was an NIF-format by group discussion interaction effect such that, only when in a peer discussion group, the NIF-formatted materials (compared to topically formatted materials) related to increased reports of perceptions of the ELSI-specific trustworthiness of policymakers (i.e., trustworthiness as related specifically to their taking into account ethical, legal, and social issues). Thus, overall, the results were inconsistent and inconclusive regarding the effects on trustworthiness.

**Conscientious Engagement**

Figure 5.2b suggested our experimental manipulations might also predict conscientious engagement (and recall there was support for this in Chap. 3, for the critical thinking manipulation). When simply examining the average correlations across studies (Table 5.1), however, we see small and somewhat unreliable correlations between conscientious engagement and the critical thinking manipulation, but more robust correlations between use of peer discussion and conscientious engagement, in all three studies where peer discussion was varied. In Study 5 there was also a significant correlation between use of strong (vs. weak) information and conscientious engagement.

Analyses regressing conscientious engagement on all of our experimental manipulations found that peer discussion increased reports of conscientious engagement in each study. However, the critical thinking manipulation was not consistently related. In Study 2, critical thinking prompts related to *reduced* conscientious engagement. Note that the reason we had revised our critical thinking prompts in Studies 3–5 was partly because of noticing the negative impact our critical thinking manipulations in Studies 1 and 2 had on engagement. In Studies 3 and 4, critical thinking prompts sometimes increased conscientious engagement, as noted in Chap. 3.

---

[7]Although active facilitation in Study 5 appeared to be positively related to perceptions of the information (as shown in Table 5.1), assignment to active (vs. passive) facilitation occurred *after* students had read and rated the background information. This suggests that our random assignment procedures were not entirely effective and inadvertently created a situation where those who saw the materials more positively (trustworthy and clear) ended up more likely to be assigned to the active moderation condition.

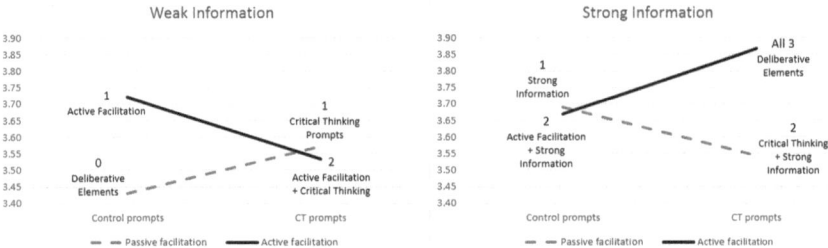

**Fig. 5.3** Illustration of the three-way interaction between information (weak vs. strong), facilitation (passive vs. active), and deliberation prompts (control vs. critical thinking) when predicting conscientious engagement in Study 5. Note: Numbers indicate how conscientious engagement is affected by including one, two, or three elements commonly associated with high-quality deliberative events

Additional analyses revealed the positive effects of the critical thinking prompts were primarily found when other elements commonly associated with good deliberation practices were missing. In Study 4 there was evidence of an interaction such that critical thinking increased average reported conscientious engagement if one was *not* in a group for discussion and group discussion increased conscientious engagement if one was *not* in the critical thinking condition. Thus, it seemed that group discussion and critical thinking prompts again compensated for the lack of the other. In Study 5, there was a significant three-way interaction between information strength, active moderation, and critical thinking prompts, which is illustrated in Fig. 5.3. This interaction also suggested that certain positive elements could compensate for the lack of other positive deliberative elements. Compared to having none of the features that are commonly associated high-quality deliberative events (i.e., no peer discussion, weak information, and control prompts rather than prompts to think critically), adding any one of those elements increased conscientious engagement (and the largest effect was seen by adding active facilitation). However, it never helped to add an element to a situation that already had an existing single element. In fact, adding critical thinking to either of the other elements seemed to reduce conscientious engagement. On the other hand, the highest levels of conscientious engagement were achieved if you had all three of the elements.

### 5.4.4   *(3) Do Our Mediators Impact Policy Acceptance/Support or Moderate the Preference-Acceptance/Support Relationship?*

As a final step in evaluating the feasibility of the models illustrated in Fig. 5.2 across our studies, we tested for the impacts of our mediators on policy acceptance/support and tested also whether the mediators might moderate the policy preference effect on policy acceptance/support. For these analyses, we conducted multiple regressions that always included a dummy code for the scenario type (pro-development or pro-regulation), as well as the effect of policy preference, and the interaction between scenario type and policy preference if relevant. Note that, prior to testing the simple main effect of a mediator and its interaction with policy preference, we tested for and if possible ruled out additional interactions with scenario type (i.e., the two-way mediator x scenario interaction and the three-way mediator x policy preference x scenario interaction). This was because prior analyses, including the correlation between scenario and policy acceptance, suggested that some effects differed depending on whether participants received the pro-development or pro-regulation policy decision scenario. Whenever we found an interaction with scenario type, we report the two-way interactions (between the mediator and policy preference) separately for each scenario type in Table 5.3.

**Process Perceptions**
Selected results from the regression analyses examining the impacts of our process perception variables predicting policy acceptance/support are shown in Table 5.3. As shown in the top section of Table 5.3, perceptions of process fairness and process competence consistently and positively related to (had main effects upon) policy acceptance in each study.

Process perceptions also often interacted with policy preferences to predict policy acceptance/support. Contrary to our expectations, when interactions emerged between policy preferences and process perceptions, the interactions were usually positive, indicating that perceptions of fairness and competence *increased* the extent to which policy preferences positively predicted acceptance. For example, Table 5.3 shows that process fairness in Study 3 had a positive impact (B = 0.45) on policy acceptance for those in the pro-regulation condition at mean levels of policy preference. The interaction between fairness and policy preference differed by scenario (see rightmost set of columns for the interactions). For those in the pro-development scenario condition, the relationship between process fairness and policy acceptance increased (grew stronger) as one's positive preferences increased (by 0.16 for every 1 SD of increase of preferences). Thus, unlike prior literature that has found procedural justice perceptions to be especially important when people receive outcomes that they do not favor, our studies relatively consistently found that the process perceptions were most highly related to policy acceptance when people received outcomes that they did prefer.

We suspect we may have found this pattern because the process perceptions in our study were assessed immediately *after* people were informed, in the scenario, of the policy decision made by the government. This outcome knowledge (in light of their policy preferences) likely impacted their perceptions of the processes used to come to the decision. Indeed, policy preferences were always strongly related to ratings of process fairness and process incompetence (see Table 5.2). We may have gotten a different pattern of results if we had assessed process perceptions prior to

**Table 5.3** Mediator variable main effects on policy acceptance/support and interactions with policy preference when predicting acceptance/support

| Mediating variable and study | Main effects | | | | Interactions with policy preferences | | | |
|---|---|---|---|---|---|---|---|---|
| | $B$ | SE | $p$ | partial Eta$^2$ | $B$ | SE | $p$ | partial Eta$^2$ |
| *Process fairness perceptions* | | | | | *Fairness interactions with policy preferences* | | | |
| Study 2 | 0.63*** | 0.07 | 0.000 | 0.356 | NS | | | |
| Study 3 (pro-reg, mean policy pref) | 0.45*** | 0.08 | 0.000 | 0.099 | −0.04 | 0.07 | 0.574 | 0.001 |
| Study 3 (pro-devel) | | | | | 0.16* | 0.08 | 0.041 | 0.016 |
| Study 4 (mean policy pref) | 0.54*** | 0.06 | 0.000 | 0.299 | 0.11+ | 0.06 | 0.072 | 0.017 |
| Study 5 | 0.32*** | 0.05 | 0.000 | 0.175 | NS | | | |
| *Process competence main effect* | | | | | *Competence interactions with policy preferences* | | | |
| Study 2 | 0.46*** | 0.07 | 0.000 | 0.217 | NS | | | |
| Study 3 (pro-reg) | 0.37*** | 0.08 | 0.000 | 0.074 | NS | | | |
| Study 3 (pro-devel) | 0.15+ | 0.08 | 0.063 | 0.013 | NS | | | |
| Study 4 (pro reg, mean policy pref) | 0.60*** | 0.08 | 0.000 | 0.220 | 0.06 | 0.07 | 0.370 | 0.004 |
| Study 4 (pro-devel) | | | | | 0.32** | 0.10 | 0.001 | 0.057 |
| Study 5 (pro reg, mean policy pref) | 0.40*** | 0.07 | 0.000 | 0.133 | −0.32*** | 0.08 | 0.000 | 0.067 |
| Study 5 (pro-devel) | | | | | −0.05 | 0.05 | 0.266 | 0.005 |
| *Biased information main effect* | | | | | *Biased information interactions with policy preferences* | | | |
| Study 2 (pro-reg, mean policy pref) | −0.18+ | 0.09 | 0.053 | 0.024 | −0.01 | 0.08 | 0.872 | 0.000 |
| Study 2 (pro-devel) | | | | | −0.32* | 0.12 | 0.010 | 0.042 |
| Study 3 | −0.06 | 0.06 | 0.295 | 0.004 | NS | | | |
| Study 4 (pro-reg, mean policy pref) | −0.17 | 0.10 | 0.082 | 0.017 | 0.05 | 0.08 | 0.531 | 0.002 |
| Study 4 (pro-devel) | | | | | −0.20+ | 0.11 | 0.076 | 0.017 |
| Study 5 | −0.07 | 0.05 | 0.175 | 0.008 | NS | | | |
| *Untrustworthy information main effect* | | | | | *Untrustworthy information interactions with policy pref.* | | | |
| Study 2 (mean policy pref) | 0.06 | 0.09 | 0.528 | 0.004 | −0.19* | 0.08 | 0.019 | 0.049 |
| Study 3 | −0.02 | 0.06 | 0.675 | 0.001 | NS | | | |
| Study 4 | −0.29** | 0.10 | 0.003 | 0.048 | NS | | | |
| Study 5 (mean policy pref) | −0.03 | 0.05 | 0.467 | 0.002 | −0.19*** | 0.05 | 0.000 | 0.061 |
| *Unclear information main effect* | | | | | *Unclear information interactions with policy preferences* | | | |
| Study 3 | −0.08 | 0.06 | 0.197 | 0.006 | NS | | | |

(continued)

**Table 5.3** (continued)

| Mediating variable and study | Main effects | | | | Interactions with policy preferences | | | |
|---|---|---|---|---|---|---|---|---|
| | $B$ | SE | $p$ | partial Eta$^2$ | $B$ | SE | $p$ | partial Eta$^2$ |
| Study 4 | −0.03 | 0.07 | 0.638 | 0.001 | NS | | | |
| Study 5 (mean policy pref) | 0.09+ | 0.05 | 0.070 | 0.015 | −0.18** | 0.05 | 0.001 | 0.050 |
| *Policymaker trustworthiness main effect* | | | | | *Policymaker trustworthiness interactions with pol. pref.* | | | |
| Study 3 | 0.08 | 0.06 | 0.188 | 0.007 | NS | | | |
| Study 4 | 0.19** | 0.07 | 0.007 | 0.038 | NS | | | |
| *Policymaker untrustworthiness main effect* | | | | | *Policymaker untrustworthiness interact. with pol. pref.* | | | |
| Study 3 | −0.02 | 0.06 | 0.725 | 0.000 | NS | | | |
| Study 4 (pro-reg, mean policy pref) | −0.05 | 0.10 | 0.586 | 0.002 | 0.19* | 0.09 | 0.032 | 0.025 |
| Study 4 (pro-devel) | | | | | −0.08 | 0.11 | 0.435 | 0.003 |
| *Policymaker ELSI-trustworth. main effect* | | | | | *Policymaker ELSI-trustworth. interactions with pol. pref.* | | | |
| Study 3 | 0.20** | 0.06 | 0.001 | 0.044 | NS | | | |
| *Conscientious engagement main effect* | | | | | *Conscientious engagement interactions with pol. pref.* | | | |
| Study 2 | −0.05 | 0.07 | 0.475 | 0.003 | NS | | | |
| Study 3 | 0.02 | 0.06 | 0.690 | 0.012 | NS | | | |
| Study 4 (mean policy pref) | −0.01 | 0.07 | 0.860 | 0.000 | 0.12+ | 0.07 | 0.086 | 0.016 |
| Study 5 (pro-reg, mean policy pref) | 0.14* | 0.06 | 0.029 | 0.021 | 0.04 | 0.06 | 0.482 | 0.002 |
| Study 5 (pro-devel) | | | | | 0.24** | 0.07 | 0.001 | 0.046 |

Note: If the main effects or interactions were dependent upon scenario type (pro-development vs. pro-regulation of nanotechnology), then multiple main effects or interactions are shown. In the case of a variable x preference x scenario-type interaction, both variable x preference interactions are reported (separately for each scenario type), but only one main effect of the variable may be reported. This main effect, if computed in the context of higher-order interations, may be conditional upon on other factors indicated in parens. NS is used to indicate when the interaction is not significant across both the pro-development and pro-regulation scenarios

+p < .10
*p < .05
**p < .01
***p < .001

revealing the policy decision outcome.[8] Future research including measures of process perceptions prior to people learning about the policy outcomes is needed to clarify these patterns.

**Information Perceptions**

As shown in Tables 5.2 and 5.3, the impacts of perceptions of information quality did not consistently have main effects on policy acceptance/support, but when effects were found, seeing the information as deficient (biased, untrustworthy, or unclear) was more likely to decrease than increase policy acceptance. Information quality perceptions also often interacted with policy preferences to predict policy acceptance/support as shown in the right-hand side of Table 5.3. Whenever the interaction occurred, it was negative, indicating that perceiving the information as inadequate reduced the effect of policy preferences on acceptance/support (and conversely, that positive quality perceptions relate to increased relationships between policy preferences and acceptance/support). As previously described and illustrated in Fig. 5.2, we thought that perceptions of high information quality might either (1) *weaken* the relationship between policy preferences and acceptance due to increasing people's procedural fairness assessments or (2) *strengthen* relationships between policy preferences and acceptance/support due to increasing attitude certainty related to their preferences. Even though we did not assess attitude certainty, the pattern we found more closely matches the second account (Figure 5.2b). In future research it would be interesting to investigate whether our effects of information perceived as poor quality are indeed due to increased uncertainty about the preferences that participants formed during the activities.          It is also noteworthy that three-way interactions with scenario type were again apparent in these analyses when examining the effect of perceptions of bias in the information. It was only in the pro-development scenario condition that increased perceptions of bias (measured at A2) related to weakened relationships between policy preferences and acceptance/support (measured at A4) (see Table 5.3, right-hand side; Study 2 pro-development condition reveals a $-0.32$ interaction effect, and Study 4 pro-development condition finds a $-0.20$ interaction effect). Note that in this case, information perceptions were assessed before learning about the outcome, making it not possible for people's reaction to the outcome to impact their perceptions of the information. While it is not entirely clear why bias perceptions would only impact preference-acceptance relationships in the pro-development condition, the three-way interactions underscore that not all policy decisions are equal and that accepting/supporting one policy might be very different than accepting/supporting a different or seemingly

---

[8] It is also worth noting that, although the three-way interaction effect was the same in all the studies where it was found, it was surprising that the patterns of two-way effects differed in Study 5 as shown in Table 5.3. Given the differences between the acceptance/support scales constructed for Study 5 compared to the other studies, we also conducted item-level analyses to conduct analyses using variables more comparable to those used in Studies 2–4. However, the same pattern of results was found as with the full scales. It is possible that differential effects were instead due to how closely tied the Study 5 acceptance/support scales were to the scenario.

opposite policy. In our context, one policy decision (pro-development) might have been viewed as more risky than the other (pro-regulation) and thus activated risk aversion and a bias toward the status quo, which may have included greater consideration of aspects of the process that led to the decision (including quality of information considerations).

**Perceptions of Policymaker Trustworthiness and Untrustworthiness**
As previously noted, measures of perceptions of policymakers were only administered to all students in Studies 3 and 4, making it more difficult to assess replication of effects or lack of effects. Nonetheless, Table 5.3 shows that, in both studies, there was at least one indication that perceived trustworthiness of policymakers has a positive main effect on policy acceptance/support. In Study 4 there was also an interaction such that perceptions of untrustworthiness increased the relationship between preferences and acceptance/support. In other words, people who perceive policymakers as untrustworthy appear to more strongly rely on their policy preferences to decide upon policy acceptance/support. Future research should investigate whether these patterns hold across additional replications. That is, are trustworthiness perceptions most important for their main effects on policy acceptance? Are (low) untrustworthiness perceptions most important for accepting policies even when they are not preferred?

**Conscientious Engagement**
Finally, there was not much evidence that conscientious engagement had a main effect on policy acceptance/support, but there was evidence in both Studies 4 and 5 that conscientious engagement impacts the policy preference-acceptance relationship. In each case, the interaction was such that those who reported they engaged more conscientiously had stronger relationships between their reported policy preferences and acceptance/support. This is consistent with the theorizing that went into Figure 5.2b, but future research will need to establish whether or not the reason for the pattern is due to increases in attitude certainty.

Once again, there was also evidence that conscientious engagement mattered most in the pro-development condition (in Study 5). Future research is needed to explain why this is so. It is possible that when people are more accepting overall of a policy (as was the case in the pro-regulation scenario), those factors such as attitude certainty matter less.

## 5.5  Summary and Conclusions

In summary, our experimental manipulations rarely had direct impacts on policy acceptance/support, and never directly moderated the relationship between policy preferences and policy acceptance/support. However, public engagement features may still matter because sometimes our experimental manipulations did impact our proposed mediators which in turn impacted policy acceptance/support or moderated the relationship between policy preferences and acceptance.

One of our most robust findings was that critical thinking prompts increased negative evaluations of the quality of information participants received. This is important because of the role of information evaluations in policy acceptance/support. That is, our analyses suggest that quality of information impacts two different and competing processes. As quality of information perceptions improve, process perceptions and perceptions of policymakers improve too, which can relate to *increased* policy acceptance.[9] At the same time, as perceptions of information improve, the relationship between policy preference and acceptance also increases, *decreasing* the extent to which people who do not get their preferred outcome will be accepting or supportive. Because our analyses rarely found main effects of information quality on policy acceptance, it is possible that these two processes cancel one another out.

There was also robust evidence that peer discussion increased reports of conscientious engagement. This could be important because conscientious engagement related to both improved perceptions of information quality and trust in policymakers—both of which, as previously noted, tend to predict greater policy acceptance.[10] In addition, conscientious engagement had a moderating effect on the policy preference-acceptance relationship, increasing that relationship in a manner similar to how the perceived quality of information increased it.

Taken together, our results suggest that certain features of public engagement that strive to meet the "deliberative ideal" will result in less acceptance of non-preferred policies. Practitioners strive to use high-quality information in deliberations and strive to have people engage by thinking carefully about the information—in a manner that is likely to increase their knowledge and their application of that knowledge to their opinions. This, however, does potentially result in people's preferences driving their support/acceptance of the policy to a greater degree than if they had not consumed high-quality information conscientiously.

Other effects of our experimental manipulations were less robust. Critical thinking prompts sometimes interacted with other factors to predict process perceptions. NIF-formatted materials sometimes interacted with other factors to predict trustworthiness of policymakers. Some effects were found in Study 5, but future research is needed to see if the effects replicate. If the findings do replicate, Study 5 results suggest that certain design features can compensate for others (such as when active moderation compensated for weak background information and vice

---

[9] Of course, it is possible that all three perceptions (of information, the engagement processes, and policymakers) could be caused by a different variable, such as positive attitudes toward the topic (nanotechnology).

[10] However, again, interest in the topic of nanotechnology could drive careful deliberation, positive attitudes toward the information, and trust in the policymakers. Additional analyses and experimental research will be needed to tease apart sequences of causation.

versa). If such compensatory effects are common in public engagement research, this may make it difficult to find effects in experiments but may also be encouraging for practitioners. That is, as they strive to do many things "right," it may be reassuring to know that not everything needs to be perfectly right to achieve benefits from deliberative engagements.

# References

Anderson, R., DeTour, J., & PytlikZillig, L. M. (2015). Deliberative public engagement procedures create more coherent and informed attitudes: Evidence from deliberations about nanotechnology. *Paper presented at the 27th Annual Convention of the Association for Psychological Science*, New York, NY.

Batel, S., Devine-Wright, P., & Tangeland, T. (2013). Social acceptance of low carbon energy and associated infrastructures: A critical discussion. *Energy Policy, 58*, 1–5.

Brockner, J. (1996). Understanding the interaction between procedural and distributive justice: The role of trust. In R. M. Kramer & T. R. Tyler (Eds.), *Trust in organizations: Frontiers of theory and research* (pp. 390–413). Thousand Oaks, CA: Sage.

Cropanzano, R., & Folger, R. (1991). Procedural justice and worker motivation. In R. M. Steers & L. W. Porter (Eds.), *Motivation and work behavior* (Vol. 5, 2nd ed., pp. 131–143). New York, NY: McGraw-Hill.

Drews, S., & van den Bergh, J. C. J. M. (2016). What explains public support for climate policies? A review of empirical and experimental studies. *Climate Policy, 16*(7), 855–876. https://doi.org/10.1080/14693062.2015.1058240. Retrieved from https://doi.org/10.1080/14693062.2015.1058240

Dreyer, S. J., Polis, H. J., & Jenkins, L. D. (2017). Changing Tides: Acceptability, support, and perceptions of tidal energy in the United States. *Energy Research & Social Science, 29*, 72–83.

Dreyer, S. J., Teisl, M. F., & McCoy, S. K. (2015). Are acceptance, support, and the factors that affect them, different? Examining perceptions of US fuel economy standards. *Transportation Research Part D: Transport and Environment, 39*, 65–75.

Dreyer, S. J., & Walker, I. (2013). Acceptance and support of the Australian carbon policy. *Social Justice Research, 26*(3), 343–362.

Goodwin, S. A., & Wu, B. T. W. (1984). Attitudinal effects of pre-decision effort and post-decision unexpected favorable fait accompli events: An experiment. In J. D. Lindquist (Ed.), *Proceedings of the 1984 academy of marketing science (AMS) annual conference* (pp. 13–17). Cham, Switzerland: Springer International Publishing.

Lind, E. A., & Tyler, T. R. (1988). *The social psychology of procedural justice*. New York, NY: Plenum Press.

Mayer, R. C., Davis, J. H., & Schoorman, F. D. (1995). An integrative model of organizational trust. *Academy of Management Review, 20*, 709–734.

Rau, I., Schweizer-Ries, P., & Hildebrandt, J. (2012). Participation: The silver bullet for the acceptance of renewable energies? In S. K. Kabisch, A. K. Kunath, P. Schweizer-Reis, & A. S. Steinführer (Eds.), *Vulnerability, risks, and complexity: Impact of global change on human habitats* (pp. 177–191). Cambridge, MA: Hogrefe Publishing.

Schade, J., & Schlag, B. (2003). Acceptability of urban transport pricing strategies. *Transportation Research Part F: Traffic Psychology and Behaviour, 6*(1), 45–61. https://doi.org/10.1016/S1369-8478(02)00046-3. Retrieved from http://www.sciencedirect.com/science/article/pii/S1369847802000463

Svenson, O. L. A., & Jakobsson, M. (2010). Creating coherence in real-life decision processes: Reasons, differentiation and consolidation. *Scandinavian Journal of Psychology, 51*(2), 93–102. https://doi.org/10.1111/j.1467-9450.2009.00739.x

Tyler, T. R., & Huo, Y. J. (2002). *Trust in the law: Encouraging public cooperation with the police and courts.* New York, NY: Russell-Sage Foundation.

Wallner, J. (2008). Legitimacy and public policy: Seeing beyond effectiveness, efficiency, and performance. *Policy Studies Journal, 36*(3), 421–443.

# Chapter 6
# Conclusion and Future Directions

**Abstract** In the preceding chapters, we have presented the need for a science of public engagement, the reasons we focused on feature-process-outcome connections relating to deliberative engagement, and the basis for our targeting nanotechnology/synthetic biology as the policy area concentration of our research. In this chapter, we briefly summarize what we have learned and offer some suggestions for future studies that will further advance the science of engagement and deliberation. We also encourage the interested reader to access our data and other supplemental files in order to conduct additional analyses of the data we collected.

**Keywords** Biology · Deliberation · Engagement features · Genomics · Nanotechnology · Public engagement · Science and technology innovations · Science of public engagement · Synthetic biology

Anticipated advances in science led to the macabre creation of life portrayed in the nineteenth-century novel by Mary Shelly, *Frankenstein*, and to the fantastical technological advances depicted in the television cartoon series "The Jetsons." These popular cultural representations illustrate how developments in science and technology both excite and frighten society, often evoking the public's interest in being involved in decisions about whether to permit, regulate, or squelch scientific and technological innovations. For example, the recent announcement of the long-awaited breakthrough in editing human genes to remedy genetic anomalies that lead to disease again raised the specter of designing babies and led to calls for public deliberation about these emerging technologies (e.g., Belluck, 2017).

As we noted in the first chapter, public engagements regarding science and technology innovations allow many in society to provide input about what is accept-

**Electronic supplementary material**: The online version of this chapter (https://doi.org/10.1007/978-3-319-78160-0_6) contains supplementary material, which is available to authorized users.

able and what is not (e.g., Delgado, Kjølberg, & Wickson, 2011). Such public involvements also can infuse public values into technology development discussions and are essential for a healthy democracy (e.g., Rip & Robinson, 2013; Wilsdon & Willis, 2004). Public engagement with science can increase the public understanding, appreciation, and opportunity to argue for or against acceptance of emerging science and technology advances (e.g., Gastil, 2017). There is, therefore, great hope for the value of public engagement.

In contrast to the great hope for engagement, there is a dearth of science about engagement. As we have argued previously (PytlikZillig & Tomkins, 2011), simply deploying listening sessions or other types of engagement with the public may not suffice: It is essential to ascertain what is a successful engagement, what works to ensure successful engagements, in what contexts, and why. As noted in Chap. 1, and as underscored by our own unwillingness to offer a hard-and-fast definition, the concept of "public engagement" itself is ill-defined. Currently, public engagement encompasses everything from opinion surveys to information campaigns, to interactive museum exhibits, to citizen science, to voting behavior, and to deliberative discussion. The definition of "public" is also broad and wide ranging. Consider, for example, that public engagement through deliberation can involve dialogues among or between peers, policymakers, technologists, scientists, and many other stakeholders. Furthermore, engagement methods and terminology used to describe those methods within studies of public engagement are widely varied; numerous dimensions of public engagement have been proposed without much consensus on which dimensions are most important to future research agendas; potential differences and opportunities for engaging marginalized populations for the most part have not been the target of theory or extended empirical focus (but see Young, 2002); and current categories of public engagement effectiveness criteria do not easily lend themselves to suggesting theories that would advance understanding of how various forms of public engagement work for different purposes and aims.

Given all these challenges, whatever is an aspiring public engagement researcher to do? Our work provides but one example of an approach forward. For our research, we functionally operationalized our engagements as deliberations about a target (learning about and assessing nanotechnology/synthetic biology) using accessible and appropriately thorough written materials as part of a class to inform students' decision-making as part of specifically designed tasks (see Chap. 2). We hope in the future much of what we need to know about public engagements we will know because experimental methods and valid assessments reveal what works to ensure engagements are successful according to clear criteria, under what circumstances, and why.

We hope that a science of public engagement will answer questions that go beyond our current data, such as whether, when, and why:

Face-to-face encounters are or are not preferable to online engagements.
Written materials are or are not a more effective way of providing background information than a brief video.
Engagement discussions are or are not more productive in small groups than in town hall formats.
Bringing people together in the real, versus the virtual, world enhances certain outcomes and so on.

In our series of studies, we were guided by affective, cognitive, and behavioral psychology to try to better understand the impacts of different *features* of engagement: Specifically, we looked at aspects of cognitive engagement (critical thinking, information organization), characteristics of background information (pro versus con perspectives of the topic, stronger versus weaker information), whether there was discussion, and active versus passive facilitation of the discussions (Chap. 2). We examined these matters in the context of college students—future scientists!—learning about the intersection of nanotechnology and genomics as part of an introductory biology course.

This sample consisted of participants who are comparatively bright and motivated and from a Midwestern, public university, so they may not generalize precisely to others across the American population. We do not think our materials always "worked" as well as our materials have when we have engaged residents on city budgeting issues. That is, in the city budgeting engagement we had both objective and subjective indications that participants learned a great deal about the way a city's finances worked and increased their trust in government after they engaged city officials about budgeting matters (Herian, Hamm, Tomkins, & PytlikZillig, 2012; PytlikZillig, Tomkins, Herian, Hamm, & Abdel-Monem, 2012; Tomkins et al., 2012; Tomkins, PytlikZillig, Herian, Abdel-Monem, & Hamm, 2010). Yet the lessons we learned from students in a much more controlled, laboratory-like setting are important first steps for beginning to understand what to do (and what not to do), despite the limitations of our program of research.

For example, we found that *reading* information related to nanogenomics had a positive impact on both objective and subjective knowledge, but *discussing* the information with other students was not important for factual knowledge gain (Chap. 3). The ways in which information was presented to students also did not make a significant difference, nor did our prompts for critical thinking directly influence knowledge. How students engaged with the nanogenomic materials they were provided impacted subjective knowledge: Students felt they learned more when they were paying closer, more conscientious, attention, when actively and metacognitively engaged with the information they received, and when thinking imaginatively about the materials. Moreover, students who were prompted to think critically and be conscientious about the science information reported less close-mindedness about the nanoscience as well as positive engagement with the materials. As a result, we found that critical thinking did in fact impact subjective knowledge through these increases in positive engagement and decreases in negative engagement. In general, we can say that our deliberative engagements, on the whole, increased knowledge, but scholars should pay closer attention to how participants cognitively engage to realize substantive knowledge gains.

Although an outcome often hoped for by deliberative theorists is increased attitude consensus, a concern that deliberation might lead to attitude polarization has been claimed, most prominently by scholars such as Cass Sunstein (e.g., Sunstein, 2000, 2002). Our analyses of the data (Chap. 4) revealed some degree of attitude change across studies but rarely in a matter that suggested polarization or extremitization of attitudes. There was some evidence of differences in extremitization when students engaged in critical thinking (versus when they did not), but these effects were not affected by whether the students engaged in discussions and usually suggested students became more *moderate* when encouraged to think critically. Further, when we manipulated the homogeneity of attitudes within groups during discussion, we did not find any differences in attitude change or extremitization in the aggregate, but we did find that this was somewhat dependent on individual-level openness. Specifically, we found some evidence that students low in openness were the most likely to exhibit extremitization in heterogeneous groups, and students' high in openness were the most likely to exhibit extremitization in homogeneous groups. Our conclusion, partly reflecting others (for a review, see Delli Carpini, Cook, & Jacobs, 2004), is that attitude change via deliberation is dependent on context as well as personality, but we did not detect evidence of polarization related to discussing the ethical and policy implications of nanoscience materials (see also Gastil, 2017; Gastil, Kahan, & Braman, 2006).

Finally, we examined the students' policy acceptance, even when government selects a policy that is inconsistent with their own preferences (Chap. 5). Again, we did not find our experimental manipulations had many direct effects on this important outcome nor did they directly moderate the relationships between policy preferences and acceptance/support. Nonetheless, sometimes our manipulations did impact potential mediators such as perceptions of the process and of the information used. These mediators and moderators ended up being important for advancing understanding of why our manipulations may not have had effects. For example, one robust finding was that critical thinking prompts led participants to perceive the information materials more negatively. Somewhat less robustly, critical thinking prompts also sometimes led to greater conscientious (careful, thorough) engagement. Interestingly, this suggests multiple competing processes can be evoked by one feature of engagement: prompting critical thinking during deliberation evokes both conscientious engagement and negative assessments of the information provided. Note that for people to accept policies they do not prefer, it is required that the typically strong relationships between policy preferences and acceptance be reduced. Our analyses found conscientious engagement tended to *strengthen* the relationship between policy preferences and acceptance, while negative assessments of the information materials were associated with *weaker* policy preference-acceptance relationships. This suggests the reason critical thinking prompts appeared to have no overall effect on the policy preference-acceptance relationship is because the prompts evoked both processes simultaneously. It also suggests that some of the features that engagement practitioners attempt to promote (conscientious thinking and high-quality information) are likely to increase preference-acceptance relationships, thereby making it more difficult rather than less difficult for those who dislike policies to accept them.

Through our multi-year research program, we learned that although it is possible to emulate some of the control features of laboratory science, the classroom does not necessarily emulate real-world deliberations environments (for a particularly interesting study of real-world deliberations, this in the legal system and the role of juries, see Gastil, Dees, Weiser, & Simmons, 2010). Research interests had to be subservient to the educational preferences, needs, and timings of the course instructors, even if they were very flexible about our use of random assignment and the content of what we gave students. Students did receive participation grades for their involvement in our activities. They also could choose to withdraw their data from our analyses; however, the vast majority did not. Still, students distinguished between core course materials and the nanogenomic information we were providing them in recitation sections, and it was clear that nanogenomics was not as important to them as other biology they needed to know for the tests they were going to take. Nonetheless, we do think that there is promise in working with science teachers to learn about what works to increase engagement with science materials, to improve science communication to non-students, and especially to increase student interest in, skill for, and willingness to think through the ethical, legal, and social implications of the science they might practice and advance in the future.

In the future, the goals of deliberative engagements with science should be clearly articulated: Do we care about increases in science knowledge (Chap. 3), social conformity versus group polarization (Chap. 4), attitude change (Chap. 4), policy acceptance (Chap. 5), feelings of fairness and opportunities to be heard (Chap. 5), science-policy consensus, and so on? Which objectives should be prioritized, and why? What role should the reality of the costs involved in preparing and executing engagement activities play in decisions about their value for these outcomes?

As the numerous references in this book reflect, there is a lot of information already available and a lot of insights that already exist. Yet an overarching science of public engagement is not as well developed or coherent as the science of fairness or trust, or the science of attitude development and change, or the science of teaching and learning, or the science of various other pertinent elements of deliberative engagements (communication, decision-making, group processes, information sharing, and so on).

So, given where we are today, how do we get to a more developed science of engagement? We believe there is great promise in conducting theory-driven, experimental studies of public engagement utilizing randomized controls. We think that other social scientists can improve on what we did in our research. In our project, we focused on future scientists deliberating about nanotechnology and synthetic biology. Programs of research on these areas are still needed, as are other important areas of science and technology, such as workplace robotics and smart and connected communities, new genetic engineering tools such as CRISPR technology, and so on. We believe deliberations are also important for outcomes we did not investigate in our studies, such as understanding and promoting justice and clarifying values inherent in policy determinations of health care, education, finance and budgeting—really, virtually any public policy area.

For those who want to make use of our data set, we have provided our methods, materials, and measures, and substantial data as part of the supplemental materials. Additional analyses beyond those we have conducted certainly are warranted. We hope our materials will be useful for training of students and provide additional insights for public engagement researchers and practitioners. Much of our data also may contain insights we did not mine. Finally, we hope lessons from our research can enhance future studies of public engagement strategies used in different contexts and for varied purposes.

The most critical takeaway we can offer is to encourage social scientists to undertake theory-driven programs of systematic research on public engagement matters. We believe our colleagues will further develop what we have started. This seems especially salient in the current sociopolitical context. As the world's resources are increasingly depleted by an ever-growing human population, it is a near certainty that scarcity, unequal distributions of resources, and survival-relevant threats will increase the cognitive biases and psychological defenses used by key actors and the publics that follow them. This in turn will make a consensus around group efforts toward a sustainable future more and more difficult to obtain. Thus, it becomes increasingly important to promote the study of methods of public engagement (including the engagement of expert, lay, policymaking, and other publics) and to examine their impacts on outcomes such as learning (which can lead to informed decisions and attitudes), well-calibrated trust among parties involved in the decisions, polarization and conflict reduction, and willingness to accept policy decisions even when those decisions may not be personally optimal or preferred.

# References

Belluck, P. (2017, August 2). In breakthrough, scientists edit a dangerous mutation from genes in human embryos. *New York Times*, available online at https://www.nytimes.com/2017/08/02/science/gene-editing-human-embryos.html?emc=edit_na_20170802&nl=breaking-news&nlid=352639&ref=cta&_r=0

Delgado, A., Kjølberg, K. L., & Wickson, F. (2011). Public engagement coming of age: From theory to practice in STS encounters with nanotechnology. *Public Understanding of Science, 20*(6), 826–845.

Delli Carpini, M. X., Cook, F. L., & Jacobs, L. R. (2004). Public deliberation, discursive participation, and citizen engagement: A review of the empirical literature. *Annual Review of Political Science, 7*, 315–344.

Gastil, J. (2017). Designing public deliberations at the intersection of science and public policy. In K. H. Jamieson, D. Kahan, & D. A. Scheufele (Eds.), *The Oxford handbook of the science of science communication* (pp. 233–242). New York, NY: Oxford University Press.

Gastil, J., Dees, E. P., Weiser, P. J., & Simmons, C. (2010). *The jury and democracy: How jury deliberations promote civic engagement and political participation.* New York, NY: Oxford University Press.

Gastil, J., Kahan, D. M., & Braman, D. (2006, March/April). Ending polarization: The good news about the culture wars. *Boston Review*, available online at http://bostonreview.net/archives/BR31.2/gastilkahanbraman.php

Herian, M. N., Hamm, J. A., Tomkins, A. J., & PytlikZillig, L. M. (2012). Public participation, procedural fairness and evaluations of local governance: The moderating role of uncertainty. *Journal of Public Administration Research and Theory, 22*, 815–840.

PytlikZillig, L. M., & Tomkins, A. J. (2011). Public engagement for informing science and technology policy: What do we know, what do we need to know, and how will we get there? *Review of Policy Research, 28*, 197–217.

PytlikZillig, L. M., Tomkins, A. J., Herian, M. N., Hamm, J. A., & Abdel-Monem, T. (2012). Public input methods impacting confidence in government. *Transforming Government: People, Process and Policy, 6*, 92–111. [Special issue, *Collaborative e-Government*, S. A. Chun, L.F. Luna-Reyes & R. Sandoval-Almazán, eds., 5–125].

Rip, A., & Robinson, D. R. (2013). Constructive technology assessment and the methodology of insertion. In N. Doorn, D. Schuurbiers, I. van de Poel, & M. E. Gorman (Eds.), *Early engagement and new technologies: Opening up the laboratory* (Vol. 16, pp. 37–53). Dordrecht: Springer Netherlands.

Sunstein, C. R. (2000). Deliberative trouble: Why groups go to extremes. *Yale Law Journal, 110,* 71–119.

Sunstein, C. R. (2002). The law of group polarization. *Journal of Political Philosophy, 10,* 175–195.

Tomkins, A. J., Hoppe, R. D., Herian, M. N., PytlikZillig, L. M., Abdel-Monem, T., & Shank, N. C. (2012). Public input for city budgeting using e-input, face-to-face discussions, and random sample surveys: The willingness of an American community to increase taxes. *Proceedings of the European Conference on e-Government* (Vol. 2), *12,* 698–707.

Tomkins, A. J., PytlikZillig, L. M., Herian, M. N., Abdel-Monem, T., & Hamm, J. A. (2010). Public input for municipal policymaking: Engagement methods and their impact on trust and confidence. In S. A. Chun, R. Sandoval, & A. Philpot (Eds.), *The proceedings of the 11th annual international conference on digital government research, public administration online: Challenges and opportunities* (pp. 41–50). New York, NY: ACM Digital Library, Digital Government Society of North America. Available online at http://portal.acm.org/citation.cfm?id=1809885&jmp=cit&coll=GUIDE&dl=GUIDE&CFID=98220794&CFTOKEN=99331109#CIT

Young, I. M. (2002). *Inclusion and democracy.* New York, NY: Oxford University Press.

Wilsdon, J., & Willis, R. (2004). *See-through science: Why public engagement needs to move upstream.* London, UK: Demos.

# Index

© The Author(s) 2018
L. M. PytlikZillig et al., *Deliberative Public Engagement with Science*,
SpringerBriefs in Psychology, https://doi.org/10.1007/978-3-319-78160-0